THE ESSENTIAL DRUCKER

THE ESSENTIAL
DRUCKER

Selections from the Management Works of

PETER F. DRUCKER

HARPERBUSINESS
An Imprint of HarperCollins*Publishers*

HarperCollins books may be purchased for educational, business, or sales promotional use. For information please write: Special Markets Department, HarperCollins Publishers Inc., 10 East 53rd Street, New York, NY 10022.

FIRST EDITION

Designed by Ellen Cipriano

Library of Congress Cataloging-in-Publication Data has been applied for.

ISBN 0-06-621087-9

01 02 03 04 05 RRD 10 9 8 7 6 5 4 3 2

CONTENTS

► III. SOCIETY

INTRODUCTION:
THE ORIGIN AND PURPOSE OF
THE ESSENTIAL DRUCKER

*T*he *Essential Drucker* is a selection from my sixty years of work and writing on management. It begins with my book *The Future of Industrial Man* (1942) and ends (so far at least) with my 1999 book *Management Challenges for the 21st Century*.

The Essential Drucker has two purposes. First, it offers, I hope, a coherent and fairly comprehensive Introduction to Management. But second, it gives an Overview of my works on management and thus answers a question that my editors and I have been asked again and again, Where do I start to read Drucker? Which of his writings are *essential*?

Atsuo Ueda, longtime Japanese friend, first conceived *The Essential Drucker*. He himself has had a distinguished career in Japanese management. And having reached the age of sixty, he recently started a second career and became the founder and chief executive officer of a new technical university in Tokyo. But for thirty years Mr. Ueda has also been my Japanese translator and editor. He has actually translated many of my books several times as they went into new Japanese editions. He is thus thoroughly familiar with my work—in fact, he knows it better than I do. As a result he increasingly got invited to conduct Japanese conferences and seminars on my work and found himself being asked over and over again—especially by younger people, both students and executives at the start of their careers—Where do I start reading Drucker?

This led Mr. Ueda to reread my entire work, to select from it the most pertinent chapters and to abridge them so that they read as if they had originally been written as *one* cohesive text. The result was a three-volume essential Drucker of fifty-seven chapters—one volume on the management of organizations; one volume on the individual in the society of organizations; one on society in general—which was published in Japan in the summer and fall of 2000 and has met with great success. It is also being published in Taiwan, mainland China and Korea, and in Argentina, Mexico, and Brazil.

It is Mr. Ueda's text that is being used for the U.S. and U.K. editions of *The Essential Drucker*. But these editions not only are less than half the size of Mr. Ueda's original Japanese version—twenty-six chapters versus the three-volumes' fifty-seven. They also have a somewhat different focus. Cass Canfield Jr. at HarperCollins in the United States—longtime friend and my U.S. editor for over thirty years—also came to the conclusion a few years ago that there was need for an introduction to, and overview of, my sixty years of management writings. But he—rightly—saw that the U.S. and U.K. (and probably altogether the Western) audience for such a work would be both broader and narrower than the audience for the Japanese venture. It would be broader because there is in the West a growing number of people who, while not themselves executives, have come to see management as an area of public interest; there are also an increasing number of students in colleges and universities who, while not necessarily management students, see an understanding of management as part of a general education; and, finally, there are a large and rapidly growing number of mid-career managers and professionals who are flocking to advanced-executive programs, both in universities and in their employing organizations. The focus would, however, also be narrower because these additional audiences need and want less an introduction to, and overview of, Drucker's work than they want a concise, comprehensive, and sharply focused Introduction to Management, and to management alone. And thus, while using Mr. Ueda's editing and abridging, Cass Canfield Jr. (with my full, indeed my enthusiastic, support) selected and edited the texts from the Japanese three-volume edition into a comprehensive, cohesive, and self-contained introduction to management—both of the management of an enterprise and of the self-management of the individual, whether executive or professional, within an enterprise and altogether in our society of managed organizations.

My readers as well as I owe to both Atsuo Ueda and Cass Canfield Jr. an enormous debt of gratitude. The two put an incredible amount of work and dedication into *The Essential Drucker*. And the end product is not only the best introduction to one's work any author could possibly have asked for. It is also, I am convinced, a truly unique, cohesive, and self-contained introduction to management, its basic principles and concerns; its problems, challenges, opportunities.

This volume, as said before, is also an overview of my works on management. Readers may therefore want to know where to go in my books to further pursue this or that topic or this or that area of particular interest to them. Here, therefore, are the sources in my books for each of twenty-six chapters of the *The Essential Drucker*:

Chapter 1 and 26 are excerpted from *The New Realities* (1988).

Chapters 2, 3, 5, 18 are excerpted from *Management, Tasks, Responsibilities, Practices* (1974).

Chapters 4 and 19 are excerpted from *Managing for the Future* (1992), and were first published in the *Harvard Business Review* (1989) and in the *Wall Street Journal* (1988), respectively.

Chapters 6, 15, and 21 are excerpted from *Management Challenges for the 21st Century* (1999).

Chapters 7 and 23 are excerpted from *Management in a Time of Great Change* (1995) and were first published in the *Harvard*

Business Review (1994) and in the *Atlantic Monthly* (1996), respectively.

Chapter 8 was excerpted from *The Practice of Management* (1954).

Chapter 9 was excerpted from *The Frontiers of Management* (1986) and was first published in the *Harvard Business Review* (1985).

Chapters 10, 11, 12, 20, 24 were excerpted from *Innovation and Entrepreneurship* (1985).

Chapters 13, 14, 16, 17 were excerpted from *The Effective Executive* (1966).

Chapters 22 and 25 were excerpted from *Post-Capitalist Society* (1993).

All these books are still in print in the United States and in many other countries.

This one-volume edition of *The Essential Drucker* does not, however, include any excerpts from five important Management books of mine: *The Future of Industrial Man* (1942); *Concept of the Corporation* (1946); *Managing for Results* (1964; the first book on what is now called "strategy," a term unknown for business forty years ago); *Managing in Turbulent Times* (1980); *Managing the Non-Profit Organization* (1990). These are important books and still widely read and used. But their subject matter is more specialized—and in some cases also more technical—than that of the books from which the chapters of the present book were chosen—and thus had to be left out of a work that calls itself Essential.

—*Peter F. Drucker*
Claremont, California
Spring 2001

I.

MANAGEMENT

1.

MANAGEMENT AS SOCIAL FUNCTION AND LIBERAL ART

When Karl Marx was beginning work on *Das Kapital* in the 1850s, the phenomenon of management was unknown. So were the enterprises that managers run. The largest manufacturing company around was a Manchester cotton mill employing fewer than three hundred people and owned by Marx's friend and collaborator Friedrich Engels. And in Engels's mill—one of the most profitable businesses of its day—there were no "managers," only "charge hands" who, themselves workers, enforced discipline over a handful of fellow "proletarians."

Rarely in human history has any institution emerged as quickly as management or had as great an impact so fast. In less than 150 years, management has transformed the social and economic fabric of the world's developed countries. It has created a global economy and set new rules for countries that would participate in that economy as equals. And it has itself been transformed. Few executives are aware of the tremendous impact management has had. Indeed, a good many are like M. Jourdain, the character in Molière's *Bourgeois Gentilhomme*, who did not know that he spoke prose. They barely realize that they practice—or mispractice—management. As a result, they are ill prepared for the tremendous challenges that now confront them. The truly important problems managers face do not come from technology or politics; they do not originate outside of management and enterprise. They are problems caused by the very success of management itself.

To be sure, the fundamental task of management remains the same: to make people capable of joint performance through common goals, common values, the right structure, and the training and development they need to perform and to respond to change. But the very meaning of this task has changed, if only because the performance of management has converted the workforce from one composed largely of unskilled laborers to one of highly educated knowledge workers.

The Origins and Development of Management

On the threshold of World War I, a few thinkers were just becoming aware of management's existence. But few people even in the most advanced countries had anything to do with it. Now the largest single group in the labor force, more than one-third of the total, are people whom the U.S. Bureau of the Census calls "managerial and professional." Management has been the main agent of this transformation. Management explains why, for the first time in human history, we can employ large numbers of knowledgeable, skilled people in productive work. No earlier society could do this. Indeed, no earlier society could support more than a handful of such people. Until quite recently, no one knew how to put people with different skills and knowledge together to achieve common goals.

Eighteenth-century China was the envy of contemporary Western intellectuals because it supplied more jobs for educated people than all of Europe did—some twenty thousand per year. Today, the United States, with about the same population China then had, graduates nearly a million college students a year, few of whom have the slightest difficulty finding well-paid employment. Management enables us to employ them.

Knowledge, especially advanced knowledge, is always specialized. By itself it produces nothing. Yet a modern business, and not only the largest ones, may employ up to ten thousand highly knowledgeable people who represent up to sixty different knowledge areas. Engineers of all sorts, designers, marketing experts, economists, statisticians, psychologists, planners, accountants, human-resources people—all working together in a joint venture. None would be effective without the managed enterprise.

There is no point in asking which came first, the educational explosion of the last one hundred years or the management that put this knowledge to productive use. Modern management and modern enterprise could not exist without the knowledge base that developed societies have built. But equally, it is management, and management alone, that makes effective all this knowledge and these knowledgeable people. The emergence of management has converted knowledge from social ornament and luxury into the true capital of any economy.

Not many business leaders could have predicted this development back in 1870, when large enterprises were first beginning to take shape. The reason was not so much lack of foresight as lack of precedent. At that time, the only large permanent organization around was the army. Not surprisingly, therefore, its command-and-control structure became the model for the men who were putting together transcontinental railroads, steel mills, modern banks, and department stores. The command model, with a very few at the top giving orders and a great many at the bottom obeying them, remained the norm for nearly one hundred years. But it was never as static as its longevity might suggest. On the contrary, it began to change almost at once, as specialized knowledge of all sorts poured into enterprise.

The first university-trained engineer in manufacturing industry was hired by Siemens in Germany in 1867—his name was Friedrich von Hefner-Alteneck. Within five years he had built a research department. Other specialized departments followed suit. By World War I the standard functions of a manufacturer had been developed: research and engineering, manufacturing, sales, finance and accounting, and a little later, human resources (or personnel).

Even more important for its impact on enterprise—and on the world economy in general—was another management-directed development that took place at this time. That was the application of management to manual work in the form of training. The child of wartime necessity, training has propelled the transformation of the world economy in the last forty years because it allows low-wage countries to do something that traditional economic theory had said could never be done: to become efficient—and yet still low-wage—competitors almost overnight.

Adam Smith reported that it took several hundred years for a country or region to develop a tradition of labor and the expertise in manual and managerial skills needed to produce and market a given product, whether cotton textiles or violins.

During World War I, however, large numbers of unskilled, preindustrial people had to be made productive workers in practically no time. To meet this need, businesses in the United States and the United Kingdom began to apply the theory of scientific management developed by Frederick W. Taylor between 1885 and 1910 to the systematic training of blue-collar workers on a large scale. They analyzed tasks and broke them down into individual, unskilled operations that could then be learned quite quickly. Further developed in World War II, training was then picked up by the Japanese and, twenty years later, by the South Koreans, who made it the basis for their countries' phenomenal development.

During the 1920s and 1930s, management was applied to many more areas and aspects of the manufacturing business. Decentralization, for instance, arose to combine the advantages of bigness and the advantages of smallness within one enterprise. Accounting went from "bookkeeping" to analysis and control. Planning grew out of the "Gantt charts" designed in 1917 and 1918 to plan war production; and so did the use of analytical logic and statistics, which employ quantification to convert experience and intuition into definitions, information, and diagnosis. Marketing evolved as a result of applying management concepts to distribution and selling. Moreover, as early as the mid-1920s and early 1930s, some American management pioneers such as Thomas Watson Sr. at the fledgling IBM; Robert E. Wood at Sears, Roebuck; and George Elton Mayo at the Harvard Business School began to question the way manufacturing was organized. They concluded that the assembly line was a short-term compromise. Despite its tremendous productivity, it was poor economics because of its inflexibility, poor use of human resources, even poor engineering. They began the thinking and experimenting that eventually led to "automation" as the way to organize the manufacturing process, and to teamwork, quality circles, and the information-based organization as the way to manage human resources. Every one of these managerial innovations represented the application of knowledge to work, the substitution of system and information for guesswork, brawn, and toil. Every one, to use Frederick Taylor's term, replaced "working harder" with "working smarter."

The powerful effect of these changes became apparent during World War II. To the very end, the Germans were by far the better strategists. Having much shorter interior lines, they needed fewer support troops and could match their opponents in combat strength. Yet the Allies won—their victory achieved by management. The United States, with one-fifth the population of all the other belligerents combined, had almost as many men in uniform. Yet it produced more war matériel than all the others taken together. It managed to transport the stuff to fighting fronts as far apart as China, Russia, India, Africa, and Western Europe. No wonder, then, that by the war's end almost all the world had become management-conscious. Or that management emerged as a recognizably distinct kind of work, one that could be studied and developed into a discipline—as happened in each country that has enjoyed economic leadership during the postwar period.

After World War II we began to see that management is not

exclusively *business* management. It pertains to every human effort that brings together in one organization people of diverse knowledge and skills. It needs to be applied to all third-sector institutions, such as hospitals, universities, churches, arts organizations, and social service agencies, which since World War II have grown faster in the United States than either business or government. For even though the need to manage volunteers or raise funds may differentiate nonprofit managers from their for-profit peers, many more of their responsibilities are the same—among them defining the right strategy and goals, developing people, measuring performance, and marketing the organization's services. *Management worldwide has become the new social function.*

Management and Entrepreneurship

One important advance in the discipline and practice of management is that both now embrace entrepreneurship and innovation. A sham fight these days pits "management" against "entrepreneurship" as adversaries, if not as mutually exclusive. That's like saying that the fingering hand and the bow hand of the violinist are "adversaries" or "mutually exclusive." Both are always needed and at the same time. And both have to be coordinated and work together. Any *existing* organization, whether a business, a church, a labor union, or a hospital, goes down fast if it does not innovate. Conversely, any *new* organization, whether a business, a church, a labor union, or a hospital, collapses if it does not manage. Not to innovate is the single largest reason for the decline of existing organizations. Not to know how to manage is the single largest reason for the failure of new ventures.

Yet few management books have paid attention to entrepreneurship and innovation. One reason is that during the period after World War II when most of those books were written, managing the existing rather than innovating the new and different was the dominant task. During this period most institutions developed along lines laid down thirty or fifty years earlier. This has now changed dramatically. We have again entered an era of innovation, and it is by no means confined to "high-tech" or to technology generally. In fact, social innovation—as this chapter tries to make clear—may be of greater importance and have much greater impact than any scientific or technical invention. Furthermore, we now have a "discipline" of entrepreneurship and innovation (see my *Innovation and Entrepreneurship*, 1986). It is clearly a part of management and rests, indeed, on well-known and tested management principles. It applies to both existing organizations and new ventures, and to both business and nonbusiness institutions, including government.

The Accountability of Management

Management books tend to focus on the function of management inside its organization. Few yet accept it as a social function. But it is precisely because management has become so pervasive as a social function that it faces its most serious challenge. To whom is management accountable? And for what? On what does management base its power? What gives it legitimacy?

These are not business questions or economic questions. They are *political* questions. Yet they underlie the most serious assault on management in its history—a far more serious assault than any mounted by Marxists or labor unions: the hostile takeover. An American phenomenon at first, it has spread throughout the non-Communist developed world. What made it possible was the emergence of the employee pension funds as the controlling shareholders of publicly owned companies. The pension funds, while legally "owners," are economically "investors"—and, indeed, often "speculators." They have no interest in the enterprise and its welfare. In fact, in the United States at least they are "trustees," and are not supposed to consider anything but immediate pecuniary gain. What underlies the takeover bid is the postulate that the enterprise's sole function is to provide the largest possible *immediate* gain to the shareholder. In the absence of any other justification for management and enterprise, the "raider" with his hostile takeover bid prevails—and only too often immediately dismantles or loots the going concern, sacrificing long-range, wealth-producing capacity to short-term gains.

Management—and not only in the business enterprise—has to be accountable for performance. But how is performance to be defined? How is it to be measured? How is it to be enforced? And to *whom* should management be accountable? That these questions can be asked is in itself a measure of the success and importance of management. That they need to be asked is, however, also an indictment of managers. They have not yet faced up to the fact that they represent power—and power has to be accountable, has to be legitimate. They have not yet faced up to the fact that they matter.

What Is Management?

But what is management? Is it a bag of techniques and tricks? A bundle of analytical tools like those taught in business schools? These are important, to be sure, just as thermometer and anatomy are important to the physician. But the evolution and history of management—its successes as well as its problems—teach that management is, above all else, based on a very few, essential principles. To be specific:

> ➤ Management is about human beings. Its task is to make people capable of joint performance, to make their strengths effective and their weaknesses irrelevant. This is what organization is all about, and it is the reason that management is the critical, determining factor. These days, practically all of us work for a managed institution, large or small, business or nonbusiness. We depend on management for our livelihoods.

And our ability to contribute to society also depends as much on the management of the organization for which we work as it does on our own skills, dedication, and effort.

➤ Because management deals with the integration of people in a common venture, it is deeply embedded in culture. What managers do in West Germany, in the United Kingdom, in the United States, in Japan, or in Brazil is exactly the same. How they do it may be quite different. Thus one of the basic challenges managers in a developing country face is to find and identify those parts of their own tradition, history, and culture that can be used as management building blocks. The difference between Japan's economic success and India's relative backwardness is largely explained by the fact that Japanese managers were able to plant imported management concepts in their own cultural soil and make them grow.

➤ Every enterprise requires commitment to common goals and shared values. Without such commitment there is no enterprise; there is only a mob. The enterprise must have simple, clear, and unifying objectives. The mission of the organization has to be clear enough and big enough to provide common vision. The goals that embody it have to be clear, public, and constantly reaffirmed. Management's first job is to think through, set, and exemplify those objectives, values, and goals.

➤ Management must also enable the enterprise and each of its members to grow and develop as needs and opportunities change. Every enterprise is a learning and teaching institution. Training and development must be built into it on all levels—training and development that never stop.

➤ Every enterprise is composed of people with different skills and knowledge doing many different kinds of work. It must be built on communication and on individual responsibility. All members need to think through what they aim to accomplish—and make sure that their associates know and understand that aim. All have to think through what they owe to others—and make sure that others understand. All have to think through what they in turn need from others—and make sure that others know what is expected of them.

➤ Neither the quantity of output nor the "bottom line" is by itself an adequate measure of the performance of management and enterprise. Market standing, innovation, productivity, development of people, quality, financial results—all are crucial to an organization's performance and to its survival. Nonprofit institutions too need measurements in a number of areas specific to their mission. Just as a human being needs a diversity of measures to assess his or her health and performance, an organization needs a diversity of measures to assess its health and performance. Performance has to be built into the enterprise and its management; it has to be measured—or at least judged—and it has to be continually improved.

➤ Finally, the single most important thing to remember about any enterprise is that results exist only on the outside. The result of a business is a satisfied customer. The result of a hospital is a healed patient. The result of a school is a student who has learned something and puts it to work ten years later. Inside an enterprise, there are only costs.

Managers who understand these principles and function in their light will be achieving, accomplished managers.

Management as a Liberal Art

Thirty years ago the English scientist and novelist C. P. Snow talked of the "two cultures" of contemporary society. Management, however, fits neither Snow's "humanist" nor his "scientist." It deals with action and application; and its test is results. This makes it a technology. But management also deals with people, their values, their growth and development—and this makes it a humanity. So does its concern with, and impact on, social structure and the community. Indeed, as everyone has learned who, like this author, has been working with managers of all kinds of institutions for long years, management is deeply involved in moral concerns—the nature of man, good and evil.

Management is thus what tradition used to call a liberal art—"liberal" because it deals with the fundamentals of knowledge, self-knowledge, wisdom, and leadership; "art" because it is also concerned with practice and application. Managers draw on all the knowledges and insights of the humanities and the social sciences—on psychology and philosophy, on economics and history, on ethics—as well as on the physical sciences. But they have to focus this knowledge on effectiveness and results—on healing a sick patient, teaching a student, building a bridge, designing and selling a "user-friendly" software program.

For these reasons, management will increasingly be the discipline and the practice through which the "humanities" will again acquire recognition, impact, and relevance.

2.

THE DIMENSIONS
OF MANAGEMENT

Business enterprises—and public-service institutions as well—are organs of society. They do not exist for their own sake, but to fulfill a specific social purpose and to satisfy a specific need of a society, a community, or individuals. They are not ends in themselves, but means. The right question to ask in respect to them is not, What are they? but, What are they supposed to be doing and what are their tasks?

Management, in turn, is the organ of the institution.

The question, What is management? comes second. First we have to define management in and through its tasks.

There are three tasks, equally important but essentially different, that management has to perform to enable the institution in its charge to function and to make its contribution.

> ➤ Establishing the specific purpose and mission of the institution, whether business enterprise, hospital, or university
> ➤ Making work productive and the worker effective
> ➤ Managing social impacts and social responsibilities

Mission

An institution exists for a specific purpose and mission; it has a specific social function. In the business enterprise, this means economic performance.

With respect to this first task, the task of economic performance, business and nonbusiness institutions differ. In respect to every other task, they are similar. But only business has economic performance as its specific mission; it is the definition of a business that it exists for the sake of economic performance. In all other institutions—hospital, church, university, or armed services—economic considerations are a restraint. In business enterprise, economic performance is the rationale and purpose.

Business management must always, in every decision and action, put economic performance first. It can justify its existence and its authority only by the economic results it produces. A business management has failed if it does not produce economic results. It has failed if it does not supply goods and services desired by the consumer at a price the consumer is willing to pay. It has failed if it does not improve, or at least maintain, the wealth-producing capacity of the economic resources entrusted to it. And this, whatever the economic or political structure or ideology of a society, means responsibility for profitability.

Worker Achievement

The second task of management is to make work productive and the worker effective. A business enterprise (or any other institution) has only one true resource: people. It succeeds by making human resources productive. It accomplishes its goals through work. To make work productive is, therefore, an essential function. But at the same time, these institutions in today's society are increasingly the means through which individual human beings find their livelihood, find their access to social status, to community and to individual achievement and satisfaction. To make the worker productive is, therefore, more and more important and is a measure of the performance of an institution. It is increasingly a task of management.

Organizing work according to its own logic is only the first step. The second and far more difficult one is making work suitable for human beings—and their logic is radically different from the logic of work. Making the worker achieving implies consideration of the human being as an organism having peculiar physiological and psychological properties, abilities, and limitations, and a distinct mode of action.

Social Responsibilities

The third task of management is managing the social impacts and the social responsibilities of the enterprise. None of our institutions exists by itself and is an end in itself. Every one is an organ of society and exists for the sake of society. Business is no exception. Free enterprise cannot be justified as being good for business; it can be justified only as being good for society.

Business exists to supply goods and services to customers, rather than to supply jobs to workers and managers, or even dividends to stockholders. The hospital does not exist for the sake of doctors and nurses, but for the sake of patients whose one and only desire is to leave the hospital cured and never come back. Psychologically, geographically, culturally, and socially, institutions must be part of the community.

To discharge its job, to produce economic goods and services, the business enterprise has to have impact on people, on communities, and on society. It has to have power and authority over people, e.g., employees, whose own ends and purposes are not defined by and within the enterprise. It has to have impact on the community as a neighbor, as the source of jobs and tax revenue (but also of waste products and pollutants). And, increasingly, in our pluralist society of organizations, it has to add to its fundamental concern for the quantities of life—i.e., economic goods and services—concern for the quality of life, that is, for the physical, human, and social environment of modern man and modern community.

3.

THE PURPOSE AND OBJECTIVES OF A BUSINESS

Asked what a business is, the typical businessman is likely to answer, "An organization to make a profit." The typical economist is likely to give the same answer. This answer is not only false, it is irrelevant.

The prevailing economic theory of the mission of business enterprise and behavior, the maximization of profit—which is simply a complicated way of phrasing the old saw of buying cheap and selling dear—may adequately explain how Richard Sears operated. But it cannot explain how Sears, Roebuck or any other business enterprise operates, or how it should operate. The concept of profit maximization is, in fact, meaningless. The danger in the concept of profit maximization is that it makes profitability appear a myth.

Profit and profitability are, however, crucial—for society even more than for the individual business. Yet profitability is not the purpose of, but a limiting factor on business enterprise and business activity. Profit is not the explanation, cause, or rationale of business behavior and business decisions, but rather the test of their validity.

If archangels instead of businessmen sat in directors' chairs, they would still have to be concerned with profitability, despite their total lack of personal interest in making profits.

The root of the confusion is the mistaken belief that the motive of a person—the so-called profit motive of the businessman—is an explanation of his behavior or his guide to right action. Whether there is such a thing as a profit motive at all is highly doubtful. The idea was invented by the classical economists to explain the economic reality that their theory of static equilibrium could not explain. There has never been any evidence for the existence of the profit motive, and we have long since found the true explanation of the phenomena of economic change and growth which the profit motive was first put forth to explain.

It is irrelevant for an understanding of business behavior, profit, and profitability, whether there is a profit motive or not. That Jim Smith is in business to make a profit concerns only him and the Recording Angel. It does not tell us what Jim Smith does and how he performs. We do not learn anything about the work of a prospector hunting for uranium in the Nevada desert by being told that he is trying to make his fortune. We do not learn anything about the work of a heart specialist by being told that he is trying to make a livelihood, or even that he is trying to benefit humanity. The profit motive and its offspring maximization of profits are just as irrelevant to the function of a business, the purpose of a business, and the job of managing a business.

In fact, the concept is worse than irrelevant: it does harm. It is a major cause of the misunderstanding of the nature of profit in our society and of the deep-seated hostility to profit, which are among the most dangerous diseases of an industrial society. It is largely responsible for the worst mistakes of public policy—in this country as well as in Western Europe—which are squarely based on the failure to understand the nature, function, and purpose of business enterprise. And it is in large part responsible for the prevailing belief that there is an inherent contradiction between profit and a company's ability to make a social contribution. Actually, a company can make a social contribution only if it is highly profitable.

To know what a business is, we have to start with its *purpose*. Its purpose must lie outside of the business itself. In fact, it must lie in society since business enterprise is an organ of society. There is only one valid definition of business purpose: *to create a customer*.

Markets are not created by God, nature, or economic forces but by businesspeople. The want a business satisfies may have been felt by the customer before he or she was offered the means of satisfying it. Like food in a famine, it may have dominated the customer's life and filled all his waking moments, but it remained a potential want until the action of businesspeople converted it into effective demand. Only then is there a customer and a market. The want may have been unfelt by the potential customer; no one knew that he wanted a Xerox machine or a computer until these became available. There may have been no want at all until business action created it—by innovation, by credit, by advertising, or by salesmanship. In every case, it is business action that creates the customer.

It is the customer who determines what a business is. It is the customer alone whose willingness to pay for a good or for a service converts economic resources into wealth, things into goods. What the customer buys and considers value is never just a product. It is always a utility, that is, what a product or service does for him.

The Purpose of a Business

Because its purpose is to create a customer, the business enterprise has two—and only these two—basic functions: marketing and innovation.

Despite the emphasis on marketing and the marketing approach, marketing is still rhetoric rather than reality in far too many businesses. "Consumerism" proves this. For what consumerism demands of business is that it actually market. It demands that business start out with the needs, the realities, the values, of the customers. It demands that business define its goal as the satisfaction of customer needs. It demands that business base its reward on its contribution to the customer. That after twenty years of marketing rhetoric consumerism could become a powerful popular movement proves that not much marketing has been practiced. Consumerism is the "shame of marketing."

But consumerism is also the opportunity of marketing. It will force businesses to become market-focused in their actions as well as in their pronouncements.

Above all, consumerism should dispel the confusion that largely explains why there has been so little real marketing. When managers speak of marketing, they usually mean the organized performance of all *selling* functions. This is still selling. It still starts out with "our products." It still looks for "our market." True marketing starts out the way Sears starts out—with the customer, his demographics, his realities, his needs, his values. It does not ask, What do we want to sell? It asks, What does the customer want to buy? It does not say, This is what our product or service does. It says, These are the satisfactions the customer looks for, values, and needs.

Indeed, selling and marketing are antithetical rather than synonymous or even complementary.

There will always, one can assume, be the need for some selling. But the aim of marketing is to make selling superfluous. The aim of marketing is to know and understand the customer so well that the product or service fits him and sells itself.

Marketing alone does not make a business enterprise. In a static economy there are no business enterprises. There are not even businesspeople. The middleman of a static society is a broker who receives his compensation in the form of a fee, or a speculator who creates no value.

A business enterprise can exist only in an expanding economy, or at least in one that considers change both natural and acceptable. And business is the specific organ of growth, expansion, and change.

The second function of a business is, therefore, *innovation*—the provision of different economic satisfactions. It is not enough for the business to provide just any economic goods and services; it must provide better and more economic ones. It is not necessary for a business to grow bigger; but it is necessary that it constantly grow better.

Innovation may result in a lower price—the datum with which the economist has been most concerned, for the simple reason that it is the only one that can be handled by quantitative tools. But the result may also be a new and better product, a new convenience, or the definition of a new want.

The most productive innovation is a *different* product or service creating a new potential of satisfaction, rather than an improvement. Typically this new and different product costs more—yet its overall effect is to make the economy more productive.

The antibiotic drug costs far more than the cold compress, which was all yesterday's physician had to fight pneumonia.

Innovation may be finding new uses for old products. A salesman who succeeds in selling refrigerators to Eskimos to prevent food from freezing would be as much of an innovator as if he had developed brand-new processes or invented a new product. To sell Eskimos a refrigerator to keep food cold is finding a new market; to sell a refrigerator to keep food from getting too cold is actually creating a new product. Technologically there is, of course, only the same old product; but economically there is innovation.

Above all, innovation is not *invention*. It is a term of economics rather than of technology. Nontechnological innovations—social or economic innovations—are at least as important as technological ones.

In the organization of the business enterprise, innovation can no more be considered a separate function than marketing. It is not confined to engineering or research but extends across all parts of the business, all functions, all activities. It cannot be confined to manufacturing. Innovation in distribution has been as important as innovation in manufacturing; and so has been innovation in an insurance company or in a bank. Innovation can be defined as the task of endowing human and material resources with new and greater wealth-producing capacity.

Managers must convert society's needs into opportunities for profitable business. That, too, is a definition of innovation. It should be stressed today, when we are so conscious of the needs of society, schools, health-care systems, cities, and environment.

Today's business enterprise (but also today's hospital or government agency) brings together a great many men of high knowledge and skill, at practically every level of the organization. But high knowledge and skill also mean decision-impact on how the work is to be done and on what work is actually being tackled.

As a result, decisions affecting the entire business and its capacity to perform are made at all levels of the organization, even fairly low ones. Risk-taking decisions—what to do and what not to do; what to continue work on and what to abandon; what products, markets, and technologies to pursue with energy and what markets, products, and technologies to ignore—are in the reality of today's business enterprise made every day by a host of people of subordinate rank, very often by people without traditional managerial title or position, e.g., research scientists, design engineers, product planners, and tax accountants.

Every one of these men and women bases their decisions on some, if only vague, theory of the business. Every one, in other words, has an answer to the question, What is our business and what should it be? Unless, therefore, the business itself—and that means its top management—has thought through the question and formulated the answer—or answers—to it, the decision-makers in the business, will decide and act on the basis of different, incompatible, and conflicting theories. They will pull in different directions without even being aware of their divergences. But they will also decide and act on the basis of wrong and misdirecting theories of the business. Common vision, common understanding, and unity of direction and effort of the entire organization require definition of "what our business is and what it should be."

Nothing may seem simpler or more obvious than to know what

a company's business is. A steel mill makes steel, a railroad runs trains to carry freight and passengers, an insurance company underwrites fire risks, and a bank lends money. Actually, What is our business? is almost always a difficult question and the right answer is usually anything but obvious.

The answer to the question, What is our business? is the first responsibility of top management.

That business purpose and business mission are so rarely given adequate thought is perhaps the single most important cause of business frustration and business failure. Conversely, in outstanding businesses such as the Telephone Company or Sears, success always rests to a large extent on raising the question, What is our business? clearly and deliberately, and on answering it thoughtfully and thoroughly.

With respect to the definition of business purpose and business mission, there is only one such focus, one starting point. It is the customer. The customer defines the business. A business is not defined by the company's name, statutes, or articles of incorporation. It is defined by the want the customer satisfies when he or she buys a product or a service. To satisfy the customer is the mission and purpose of every business. The question, What is our business? can, therefore, be answered only by looking at the business from the outside, from the point of view of customer and market. All the customer is interested in are his or her own values, wants, and reality. For this reason alone, any serious attempt to state "what our business is" must start with the customer's realities, his situation, his behavior, his expectations, and his values.

Who is the customer? is thus the first and most crucial question to be asked in defining business purpose and business mission. It is not an easy, let alone an obvious, question. How it is being answered determines, in large measure, how the business defines itself.

The consumer—that is, the ultimate user of a product or a service—is always a customer. But there is never *the* customer; there are usually at least two—sometimes more. Each customer defines a different business, has different expectations and values, buys something different.

Most businesses have at least two customers. The rug and carpet industry has both the contractor and the homeowner for its customers. Both have to buy if there is to be a sale. The manufacturers of branded consumer goods always have two customers at the very least: the housewife and the grocer. It does not do much good to have the housewife eager to buy if the grocer does not stock the brand. Conversely, it does not do much good to have the grocer display merchandise advantageously and give it shelf space if the housewife does not buy.

It is also important to ask, Where is the customer? One of the secrets of Sears's success in the 1920s was the discovery that its old customer was now in a different place: the farmer had become mobile and was beginning to buy in town.

The next question is, What does the customer buy?

The Cadillac people say that they make an automobile, and their business is called the Cadillac Motor Division of General Motors. But does the man who buys a new Cadillac buy transportation, or does he buy primarily prestige? Does the Cadillac compete with Chevrolet, Ford, and Volkswagen? Nicholas Dreystadt, the German-born service mechanic who took over Cadillac in the Great Depression years of the 1930s, answered: "Cadillac competes with diamonds and mink coats. The Cadillac customer does not buy 'transportation' but 'status.'" This answer saved Cadillac, which was about to go under. Within two years or so, it made it into a major growth business despite the depression.

Most managements, if they ask the question at all, ask, What is our business? when the company is in trouble. Of course, then it *must* be asked. And then asking the question may, indeed, have spectacular results and may even reverse what appears to be irreversible decline. To wait until a business—or an industry—is in trouble is playing Russian roulette. It is irresponsible management. The question should be asked at the inception of a business—and particularly for a business that has ambitions to grow. The most important time to ask seriously, What is our business? is when a company has been successful.

Success always makes obsolete the very behavior that achieved it. It always creates new realities. It always creates, above all, its own and different problems. Only the fairy tale ends, "They lived happily ever after."

It is not easy for the management of a successful company to ask, What is our business? Everybody in the company then thinks that the answer is so obvious as not to deserve discussion. It is never popular to argue with success, never popular to rock the boat.

Sooner or later even the most successful answer to the question, What is our business? becomes obsolete. Very few definitions of the purpose and mission of a business have anything like a life expectancy of thirty, let alone fifty, years. To be good for ten years is probably all one can normally expect.

In asking, What is our business? management therefore also needs to add, And what *will* it be? What changes in the environment are already discernible that are likely to have high impact on the characteristics, mission, and purpose of our business? and How do we *now* build these anticipations into our theory of the business, into its objectives, strategies, and work assignments?

Again the market, its potential and its trends, is the starting point. How large a market can we project for our business in five or ten years—assuming no basic changes in customers, in market structure, or in technology? And, what factors could validate or disprove those projections?

The most important of these trends is one to which few businesses pay much attention: changes in population structure and population dynamics. Traditionally businessmen, following the economists, have assumed that demographics are a constant. Historically this has been a sound assumption. Populations used to change very slowly except as a result of catastrophic events, such as major war or famine. This is no longer true, however. Populations nowadays can and do change drastically, in developed as well as in developing countries.

The importance of demographics does not lie only in the impact population structure has on buying power and buying habits, and on the size and structure of the workforce. Population shifts are the only events regarding the future for which true prediction is possible.

Management needs to anticipate changes in market structure resulting from changes in the economy, from changes in fashion or taste, and from moves by competition. And competition must always be defined according to the customer's concept of what product or service he buys and thus must include indirect as well as direct competition.

Finally, management has to ask which of the consumer's wants are not adequately satisfied by the products or services offered him today. The ability to ask this question and to answer it correctly usually makes the difference between a growth company and one that depends for its development on the rising tide of the economy or of the industry. But whoever is content to rise with the tide will also fall with it.

What Should Our Business Be?

What *will* our business be? aims at adaptation to anticipated changes. It aims at modifying, extending, and developing the existing, ongoing business.

But there is need also to ask, What *should* our business be? What opportunities are opening up or can be created to fulfill the purpose and mission of the business by making it into a *different* business?

Businesses that fail to ask this question are likely to miss their major opportunity.

Next to changes in society, economy, and market as factors demanding consideration in answering the question What should our business be? comes, of course, innovation, one's own and that of others.

Just as important as the decision on what new and different

things to do is planned, systematic abandonment of the old that no longer fits the purpose and mission of the business, no longer conveys satisfaction to the customer or customers, no longer makes a superior contribution.

An essential step in deciding what our business is, what it will be, and what it should be is, therefore, systematic analysis of all existing products, services, processes, markets, end uses, and distribution channels. Are they still viable? And are they likely to remain viable? Do they still give value to the customer? And are they likely to do so tomorrow? Do they still fit the realities of population and markets, of technology and economy? And if not, how can we best abandon them—or at least stop pouring in further resources and efforts? Unless these questions are being asked seriously and systematically, and unless managements are willing to act on the answers to them, the best definition of "what our business is, will be, and should be," will remain a pious platitude. Energy will be used up in defending yesterday. No one will have the time, resources, or will to work on exploiting today, let alone to work on making tomorrow.

Defining the purpose and mission of the business is difficult, painful, and risky. But it alone enables a business to set objectives, to develop strategies, to concentrate its resources, and to go to work. It alone enables a business to be managed for performance.

The basic definitions of the business, and of its purpose and mission, have to be translated into objectives. Otherwise, they remain insights, good intentions, and brilliant epigrams that never become achievement.

1. Objectives must be derived from "what our business is, what it will be, and what it should be." They are not abstractions. They are the action commitments through which the mission of a business is to be carried out, and the standards against which performance is to be measured. Objectives, in other words, represent the *fundamental strategy of a business.*

2. Objectives must be *operational.* They must be capable of

being converted into specific targets and specific assignments. They must be capable of becoming the basis, as well as the motivation, for work and achievement.

3. Objectives must make possible *concentration* of resources and efforts. They must winnow out the fundamentals among the goals of a business so that the key resources of men, money, and physical facilities can be concentrated. They must, therefore, be selective rather than encompass everything.

4. There must be *multiple objectives* rather than a single objective.

Much of today's lively discussion of management by objectives is concerned with the search for the "one right objective." This search is not only likely to be as unproductive as the quest for the philosophers' stone; it does harm and misdirects. To manage a business is to balance a variety of needs and goals. And this requires multiple objectives.

5. Objectives are needed in all areas on which the *survival* of the business depends. The specific targets, the goals in any area of objectives, depend on the strategy of the individual business. But the areas in which objectives are needed are the same for all businesses, for all businesses depend on the same factors for their survival.

A business must first be able to create a customer. There is, therefore, need for a *marketing objective*. Businesses must be able to innovate or else their competitors will render them obsolete. There is need for an *innovation objective*. All businesses depend on the three factors of production of the economist, that is, on *human resources, capital resources,* and *physical resources*. There must be objectives for their supply, their employment, and their development. The resources must be employed productively and their productivity has to grow if the business is to survive. There is need, therefore, for *productivity objectives*. Business exists in society and community and, therefore, has to discharge social responsibilities, at least to the point where it takes responsibility for its impact upon the environment. Therefore, objectives in respect to the *social dimensions* of business are needed.

Finally, there is need for *profit*—otherwise none of the objectives can be attained. They all require effort, that is, cost. And they can be financed only out of the profits of a business. They all entail risks; they all, therefore, require a profit to cover the risk of potential losses. Profit is not an objective but it is a requirement that has to be objectively determined in respect to the individual business, its strategy, its needs, and its risks.

Objectives, therefore, have to be set in these eight key areas:

- Marketing
- Innovation
- Human resources
- Financial resources
- Physical resources
- Productivity
- Social responsibility
- Profit requirements

Objectives are the basis for work and assignments.

They determine the structure of the business, the key activities that must be discharged, and, above all, the allocation of people to tasks. Objectives are the foundation for designing both the structure of the business and the work of individual units and individual managers.

Objectives are always needed in all eight key areas. The area without specific objectives will be neglected. Unless we determine what will be measured and what the yardstick of measurement in an area will be, the area itself will not be seen.

The measurements available for the key areas of a business enterprise are still haphazard by and large. We do not even have adequate concepts, let alone measurements, except for market standing. For something as central as profitability, we have only a rubber yardstick; and we have no real tools at all to determine how much profitability is necessary. In respect to innovation and, even more, to productivity, we hardly know more than that something ought to be done. In the other areas—including physical and financial resources—we are reduced to statements of intentions; we do not possess goals and measurements for their attainment.

However, enough is known about each area to give a progress report at least. Enough is known for each business to go to work on objectives.

We know one more thing about objectives: how to use them.

If objectives are only good intentions, they are worthless. They must be transformed into work. And work is always specific, always has—or should have—clear, unambiguous, measurable results, a deadline and a specific assignment of accountability.

But objectives that become a straitjacket do harm. Objectives are always based on expectations. And expectations are, at best, informed guesses. Objectives express an appraisal of factors that are largely outside the business and not under its control. The world does not stand still.

The proper way to use objectives is the way an airline uses schedules and flight plans. The schedule provides for the 9:00 A.M. flight from Los Angeles to get to Boston by 5:00 P.M. But if there is a blizzard in Boston that day, the plane will land in Pittsburgh instead and wait out the storm. The flight plan provides for flying at thirty thousand feet and for flying over Denver and Chicago. But if the pilot encounters turbulence or strong headwinds, he will ask flight control for permission to go up another five thousand feet and to take the Minneapolis–Montreal route. Yet no flight is ever operated without schedule and flight plan. Any change is immediately fed back to produce a new schedule and flight plan. Unless 97 percent or so of its flights proceed on the original schedule and flight plan—or within a very limited range of deviation from either—a well-run airline gets another operations manager who knows his job.

Objectives are not fate; they are directions. They are not commands; they are commitments. They do not determine the future; they are means to mobilize the resources and energies of the business for the making of the future.

Marketing Objectives

Marketing and innovation are the foundation areas in objective setting. It is in these two areas that a business obtains its results. It is performance and contribution in these areas for which a customer pays.

It is somewhat misleading to speak of a marketing objective. Marketing performance requires a number of objectives. For example, it is geared toward:

> ➤ Existing products and services in existing and present markets
> ➤ Abandonment of "yesterday" in product, services, and markets
> ➤ New products and services for existing markets
> ➤ New markets
> ➤ The distributive organization
> ➤ Service standards and services performance
> ➤ Credit standards and credit performance

Many books have been written about every one of these areas. But it is almost never stressed that objectives in these areas can be set only after two key decisions have been made: the decision on concentration, and the decision on market standing.

Archimedes, one of the great scientists of antiquity, is reported to have said; "Give me a place to stand on, and I can lift the universe off its hinges." The place to stand on is the area of concentration. It is the area that gives a business the leverage that lifts the universe off its hinges. The concentration decision is, therefore, crucial; it converts, in large measure, the definition of "what our business is" into meaningful operational commitment.

The other major decision underlying marketing objectives is that on market standing. One common approach is to say, We want to be the leader. The other one is to say, We don't care what share of the market we have as long as sales go up. Both sound plausible, but both are wrong.

Obviously, not everybody can be the leader. One has to decide in which segment of the market, with what product, what services, what values, one should be the leader. It does not do much good for a company's sales to go up if it loses market share, that is, if the market expands much faster than the company's sales do.

A company with a small share of the market will eventually become marginal in the marketplace, and thereby exceedingly vulnerable.

Market standing, regardless of the sales curve, is therefore essential. The point at which the supplier becomes marginal varies from industry to industry. But to be a marginal producer is dangerous for long-term survival.

There is also a maximum market standing above which it may be unwise to go—even if there were no antitrust laws. Market domination tends to lull the leader to sleep; monopolists flounder on their own complacency rather than on public opposition. Market domination produces tremendous internal resistance against any innovation and thus makes adaptation to change dangerously difficult.

There is also well-founded resistance in the marketplace to dependence on one dominant supplier. Whether it is the purchasing agent of a manufacturing company, the procurement officer in the air force, or the housewife, no one likes to be at the mercy of the monopoly supplier.

Finally, the dominant supplier in a rapidly expanding, especially a new, market is likely to do less well than if it shared that market with one or two other major and competing suppliers. This may seem paradoxical—and most businesspeople find it difficult to accept. But the fact is that a new market, especially a new major market, tends to expand much more rapidly when there are several suppliers rather than only one. It may be very flattering to a supplier's ego to have 80 percent of a market. But if as a result of domination by a single source, the market does not expand as it otherwise might, the supplier's revenues and profits are likely to be considerably lower than they would be if two suppliers shared a fast-expanding market. Eighty percent of 100 is considerably less than 50 percent of 250. A new market that has only one supplier is likely to become static at 100. It will be limited by the imagination of the one supplier who always knows what his product or service cannot or should not be used for. If there are several suppliers, they are likely to uncover and promote markets and end uses the single supplier never dreams of. And the market might grow rapidly to 250.

Du Pont seems to have grasped this. In its most successful innovations, Du Pont retains a sole-supplier position only until the new product has paid for the original investment. Then Du Pont licenses the innovation and launches competitors deliberately. As a result, a number of aggressive companies start developing new markets and new uses for the product. Nylon would surely have grown much more slowly without Du Pont–sponsored competition. Its markets are still growing, but without competition it would probably have begun to decline in the early 1950s, when newer synthetic fibers were brought on the market by Monsanto and Union Carbide in the United States, by Imperial Chemicals in Great Britain, and by AKU in Holland.

The market standing to aim at is not the maximum but the *optimum.*

Innovation Objective

The innovation objective is the objective through which a company makes operational its definition of "what our business should be."

There are essentially three kinds of innovation in every business: innovation in product or service; innovation in the marketplace and consumer behavior and values; and innovation in the various skills and activities needed to make the products and services and to bring them to market. They might be called respectively product innovation, social innovation, and managerial innovation.

The problem in setting innovation objectives is measuring the relative impact and importance of various innovations. But how are we to determine what weighs more: a hundred minor but immediately applicable improvements in packaging a product, or one fundamental chemical discovery that after ten more years of hard work may change the character of the business altogether? A department store and a pharmaceutical company will answer this question differently; but so may two different pharmaceutical companies.

Resources Objectives

A group of objectives deals with the resources a business needs in order to be able to perform, with their supply, their utilization, and their productivity.

All economic activity, economists have told us for two hundred years, requires three kinds of resources: land, that is, products of nature; labor, that is, human resources; and capital, that is, the means to invest in tomorrow. The business must be able to attract all three and to put them to productive use. A business that cannot attract the people and the capital it needs will not last long.

The first sign of decline of an industry is loss of appeal to qualified, able, and ambitious people. The decline of the American railroads, for instance, did not begin after World War II—it only became obvious and irreversible then. The decline actually set in around the time of World War I. Before World War I, able graduates of American engineering schools looked for a railroad career. From the end of World War I on—for whatever reason—the railroads no longer appealed to young engineering graduates, or to any educated young people.

In the two areas of people and capital supply, genuine marketing objectives are therefore required. The key questions are: What do our jobs have to be to attract and hold the kind of people we need and want? What is the supply available on the job market? And, what do we have to do to attract it? Similarly, What does the investment in our business have to be, in the form of bank loans, long-term debts or equity, to attract and hold the capital we need?

Resource objectives have to be set in a double process. One starting point is the anticipated needs of the business, which then have to be projected on the outside, that is, on the market for land, labor, and capital. But the other starting point is these "markets" themselves, which then have to be projected onto the structure, the direction, the plans of the business.

Productivity Objectives

Attracting resources and putting them to work is only the beginning. The task of a business is to make resources productive. Every business, therefore, needs productivity objectives with respect to each of the three major resources, land, labor, and capital; and with respect to overall productivity itself.

A productivity measurement is the best yardstick for comparing managements of different units within an enterprise, and for comparing managements of different enterprises.

All businesses have access to pretty much the same resources. Except for the rare monopoly situation, the only thing that differentiates one business from another in any given field is the quality of its management on all levels. The first measurement of this crucial factor is productivity, that is, the degree to which resources are utilized and their yield.

The continual improvement of productivity is one of management's most important jobs. It is also one of the most difficult; for productivity is a balance among a diversity of factors, few of which are easily definable or clearly measurable.

Labor is only one of the three factors of production. And if productivity of labor is accomplished by making the other resources less productive, there is actually loss of productivity.

Productivity is a difficult concept, but it is central. Without productivity objectives, a business does not have direction. Without productivity measurements, it does not have control.

The Social Responsibilities Objectives

Only a few years ago managers as well as economists considered the social dimension so intangible that performance objectives could not be set. We have now learned that the intangible can become very tangible indeed. Lessons we have learned from the rise of consumerism, or from the attacks on industry for the destruction of the environment, are expensive ways for us to realize that business needs to think through its impacts and its responsibilities and to set objectives for both.

The social dimension is a survival dimension. The enterprise exists in a society and an economy. Within an institution one always tends to assume that the institution exists in a vacuum. And managers inevitably look at their business from the inside. But the business enterprise is a creature of a society and an economy, and society or economy can put any business out of existence overnight. The enterprise exists on sufferance and exists only as long as the society and the economy believe that it does a necessary, useful, and productive job.

That such objectives need to be built into the strategy of a business, rather than merely be statements of good intentions, needs to be stressed here. These are objectives that are needed not because the manager has a responsibility to society. They are needed because the manager has a responsibility to the enterprise.

Profit as a Need and a Limitation

Only after the objectives in the above key areas have been thought through and established can a business tackle the question, How much profitability do we need? To attain any of the objectives entails high risks. It requires effort, and that means cost. Profit is, therefore, needed to pay for attainment of the objectives of the business. Profit is a condition of survival. It is the cost of the future, the cost of staying in business.

A business that obtains enough profit to satisfy its objectives in the key areas is a business that has the means of survival. A business that falls short of the profitability demands made by its key objectives is a marginal and endangered business.

Profit planning is necessary. But it is planning for a needed minimum profitability rather than for that meaningless shibboleth "profit maximization." The minimum needed may well turn out to be a good deal higher than the profit goals of many companies, let alone their actual profit results.

4.

WHAT THE NONPROFITS ARE
TEACHING BUSINESS

The Girl Scouts, the Red Cross, the pastoral churches—our nonprofit organizations—are becoming America's management leaders. In two areas, strategy and the effectiveness of the board, they are practicing what most American businesses only preach. And in the most crucial area—the motivation and productivity of knowledge workers—they are truly pioneers, working out the policies and practices that business will have to learn tomorrow.

Few people are aware that the nonprofit sector is by far America's largest employer. Every other adult—a total of 80 million–plus people—works as a volunteer, giving on average nearly five hours each week to one or several nonprofit organizations. This is equal to 10 million full-time jobs. Were volunteers paid, their wages, even at minimum rate, would amount to some $150 billion, or 5 percent of GNP. And volunteer work is changing fast. To be sure, what many do requires little skill or judgment: collecting in the neighborhood for the Community Chest one Saturday afternoon a year, chaperoning youngsters selling Girl Scout cookies door to door, driving old people to the doctor. But more and more volunteers are becoming "unpaid staff," taking over the professional and managerial tasks in their organizations.

Not all nonprofits have been doing well, of course. A good many community hospitals are in dire straits. Traditional churches and synagogues of all persuasions—liberal, conservative, evangelical, fundamentalist—are still steadily losing members. Indeed, the sector overall has not expanded in the last ten to fifteen years, either in terms of the money it raises (when adjusted for inflation) or in the number of volunteers. Yet in its productivity, in the scope of its work, and in its contribution to American society, the nonprofit sector has grown tremendously in the last two decades.

The Salvation Army is an example. People convicted to their first prison term in Florida, mostly very poor black or Hispanic youths, are now paroled into the Salvation Army's custody—about twenty-five thousand each year. Statistics show that if these young men and women go to jail, the majority will become habitual criminals. But the Salvation Army has been able to rehabilitate 80 percent of them through a strict work program run largely by volunteers. And the program costs a fraction of what it would to keep the offenders behind bars.

A Commitment to Management

Underlying this program and many other effective nonprofit endeavors is a commitment to management. Twenty years ago management was a dirty word for those involved in nonprofit organizations. It meant business, and nonprofits prided themselves on being free of the taint of commercialism and above such sordid considerations as the bottom line. Now most of them have learned that nonprofits need management even more than business does, precisely because they lack the discipline of the bottom line. The nonprofits are, of course, still dedicated to "doing good." But they also realize that good intentions are no substitute for organization and leadership, for accountability, performance, and results. Those things require management and that, in turn, begins with the organization's mission.

Starting with the mission and its requirements may be the first lesson business can learn from successful nonprofits. It focuses the organization on action. It defines the specific strategies needed to attain the crucial goals. It creates a disciplined organization. It alone can prevent the most common degenerative disease of organizations, especially large ones: fragmenting their always limited resources on things that are "interesting" or "look profitable" rather than concentrating them on a very small number of productive efforts.

The best nonprofits devote a great deal of thought to defining their organization's mission. They avoid sweeping statements full of good intentions and focus, instead, on objectives that have clear-cut implications for the work their members—staff and volunteers—perform. The Salvation Army's goal, for example, is to turn society's rejects—alcoholics, criminals, derelicts—into citizens. The Girl Scouts help youngsters become confident, capable young women who respect themselves and other people. The Nature Conservancy preserves the diversity of nature's fauna and flora. Nonprofits also start with the environment, the community, the "customers" to be; they do not, as American businesses tend to do, start with the inside, that is, with the organization or with financial returns.

Willowcreek Community Church in South Barrington, Illinois, outside Chicago, has become the nation's largest church—some thirteen thousand parishioners. Yet it is barely fifteen years old. Bill Hybels, in his early twenties when he founded the church, chose the community because it had relatively few churchgoers, though the population was growing fast and churches were plentiful. He went from door to door asking, "Why don't you go to church?" Then he designed a church to answer the potential customers' needs: for instance, it offers full services on Wednesday evenings because many working parents need Sunday to spend with their children. Moreover, Hybels continues to listen and react. The pastor's sermon is taped while it is being delivered and instantly reproduced so that parishioners can pick up a cassette when they leave the building because he was told again and again, "I need to listen when I drive home or drive to work so that I can build the message into my life." But he was also told, "The sermon always tells me to change my life but never how to do it." So now every one of Hybels's sermons ends with specific action recommendations.

A well-defined mission serves as a constant reminder of the need to look outside the organization not only for "customers" but also for measures of success. The temptation to content oneself with the "goodness of our cause"—and thus to substitute good intentions for results—always exists in nonprofit organizations. It is precisely because of this that the successful nonprofits have learned to define clearly what changes *outside* the organization constitute "results" and to focus on them.

The experience of one large Catholic hospital chain in the Southwest shows how productive a clear sense of mission and a focus on results can be. Despite the sharp cuts in Medicare payments and hospital stays during the past eight years, this chain has increased revenues by 15 percent (thereby managing to break even) while greatly expanding its services and raising both patient-care and medical standards. It has done so because the nun who is the CEO understood that she and her staff are in the business of delivering health care (especially to the poor), not running hospitals.

As a result, when health-care delivery began moving out of hospitals for medical rather than economic reasons about ten years ago, the chain promoted the trend instead of fighting it. It founded ambulatory surgery centers, rehabilitation centers, X-ray and lab networks, HMOs, and so on. The chain's motto was, If it's in the patient's interest, we have to promote it; it's then our job to make it pay. Paradoxically, the policy has filled the chain's hospitals; the freestanding facilities are so popular they generate a steady stream of referrals.

This is, of course, not so different from the marketing strategy of successful Japanese companies. But it is very different indeed from the way most Western businesses think and operate. And the difference is that the Catholic nuns—and the Japanese—start with the mission rather than with their own rewards, and with what they have to make happen outside themselves, in the marketplace, to deserve a reward.

Finally, a clearly defined mission will foster innovative ideas and help others understand why they need to be implemented—however much they fly in the face of tradition. To illustrate, consider the Daisy Scouts, a program for five-year-olds that the Girl Scouts initiated a few years back. For seventy-five years, first grade had been the minimum age for entry into a Brownie troop, and many Girl Scout councils wanted to keep it that way. Others, however, looked at demographics and saw the growing number of working women with "latchkey" kids. They also looked at the children and realized that they were far more sophisticated than their predecessors a generation ago (largely thanks to TV).

Today the Daisy Scouts are one hundred thousand strong and growing fast. It is by far the most successful of the many programs for kindergartners that have been started these last twenty years, and far more successful than any of the very expensive government programs. Moreover, it is so far the only program that has seen these critical demographic changes and children's exposure to long hours of TV viewing as an opportunity.

Effective Use of the Board

Many nonprofits now have what is still the exception in business— a functioning board. They also have something even rarer: a CEO who is clearly accountable to the board and whose performance is reviewed annually by a board committee. And they have what is rarer still: a board whose performance is reviewed annually against preset performance objectives. Effective use of the board is thus a second area in which business can learn from the nonprofit sector.

In U.S. law, the board of directors is still considered the "managing" organ of the corporation. Management authors and scholars agree that strong boards are essential and have been writing to that effect for more than twenty years. Nevertheless, the top managements of our large companies have been whittling away at the directors' role, power, and independence for more than half a century. In every single business failure of a large company in the last few decades, the board was the last to realize that things were going wrong. To find a truly effective board, you are much better advised to look in the nonprofit sector than in our public corporations.

In part, this difference is a product of history. Traditionally, the board has run the shop in nonprofit organizations—or tried to. In fact, it is only because nonprofits have grown too big and complex to be run by part-time outsiders, meeting for three hours a month, that so many have shifted to professional management. The American Red Cross is probably the largest nongovernmental agency in the world and certainly one of the most complex. It is responsible for worldwide disaster relief; it runs thousands of blood banks as well as the bone and skin banks in hospitals; it conducts training in cardiac and respiratory rescue nationwide; and it gives first-aid courses in thousands of schools. Yet it did not have a paid chief executive until 1950, and its first professional CEO came only with the Reagan era.

But however common professional management becomes—and professional CEOs are now found in most nonprofits and all the bigger ones—nonprofit boards cannot, as a rule, be rendered impotent the way so many business boards have been. No matter how much nonprofit CEOs would welcome it—and quite a few surely would—nonprofit boards cannot become their rubber stamp. Money is one reason. Few directors in publicly held corporations are substantial shareholders, whereas directors on nonprofit boards very often contribute large sums themselves, and are expected to bring in donors as well. But also, nonprofit directors tend to have a personal commitment to the organization's cause. Few people sit on a church vestry or on a school board unless they deeply care about religion or education. Moreover, nonprofit board members typically have served as volunteers themselves for a good many years and are deeply knowledgeable about the organization, unlike outside directors in a business.

Precisely because the nonprofit board is so committed and active, its relationship with the CEO tends to be highly contentious and full of potential for friction. Nonprofit CEOs complain that their board "meddles." The directors, in turn, complain that management "usurps" the board's function. This has forced an increasing number of nonprofits to realize that neither board nor CEO is "the boss." They are colleagues, working for the same goal but each having a different task. And they have learned that it is the CEO's responsibility to define the tasks of each, the board's and his or her own.

The key to making a board effective, as this example suggests, is not to talk about its function but to organize its work. More and more nonprofits are doing just that, among them half a dozen fair-sized liberal arts colleges, a leading theological seminary, and some large research hospitals and museums.

The weakening of the large corporation's board would, many of us predicted, weaken management rather than strengthen it. It would diffuse management's accountability for performance and results; and indeed, it is the rare big-company board that reviews the CEO's performance against preset business objectives. Weakening the board would also, we predicted, deprive top management of effective and credible support if it were attacked. These predictions have been borne out amply in the recent rash of hostile takeovers.

To restore management's ability to manage, we will have to make boards effective again—and that should be considered a responsibility of the CEO. A few first steps have been taken. The audit committee in most companies now has a real rather than a make-believe job responsibility. A few companies—though so far almost no large ones—have a small board committee on succession and executive development, which regularly meets with senior executives to discuss their performance and their plans. But I know of no company so far where there are work plans for the board and any kind of review of the board's performance. And few do what the larger nonprofits now do routinely: put a new board member through systematic training.

To Offer Meaningful Achievement

Nonprofits used to say, We don't pay volunteers so we cannot make demands upon them. Now they are more likely to say, Volunteers must get far greater satisfaction from their accomplishments and make a greater contribution precisely because they do not get a paycheck. The steady transformation of the volunteer from well-meaning amateur to trained, professional, unpaid staff member is the most significant development in the nonprofit sector—as well as the one with the farthest-reaching implications for tomorrow's business.

A midwestern Catholic diocese may have come furthest in this process. It now has fewer than half the priests and nuns it had only fifteen years ago. Yet it has greatly expanded its activities—in some cases, such as providing help for the homeless and for drug abusers, more than doubling them. It still has many traditional volunteers like the Altar Guild members who arrange flowers. But now it is also being served by some two thousand part-time unpaid staff who run the Catholic charities, perform administrative jobs in parochial schools, and organize youth activities, college Newman Clubs, and even some retreats.

A similar change has taken place at the First Baptist Church in Richmond, Virginia, one of the largest and oldest churches in the Southern Baptist Convention. When Dr. Peter James Flamming took over, the church had been going downhill for many years, as is typical of old, inner-city churches. Today it again has four thousand communicants and runs a dozen community outreach programs as well as a full complement of in-church ministries. The church has only nine paid full-time employees. But of its four thousand communicants, one thousand serve as unpaid staff.

This development is by no means confined to religious organizations. The American Heart Association has chapters in every city of any size throughout the country. Yet its paid staff is limited to those at national headquarters, with just a few traveling troubleshooters serving the field. Volunteers manage and staff the chapters, with full responsibility for community health education as well as fund-raising.

These changes are, in part, a response to need. With close to half the adult population already serving as volunteers, their overall number is unlikely to grow. And with money always in short supply, the nonprofits cannot add paid staff. If they want to add to their activities—and needs are growing—they have to make volunteers more productive, have to give them more work and more responsibility. But the major impetus for the change in the volunteer's role has come from the volunteers themselves.

More and more volunteers are educated people in managerial or professional jobs—some preretirement men and women in their fifties, even more baby boomers who are reaching their mid-thirties or forties. These people are not satisfied with being helpers. They are knowledge workers in the jobs in which they earn their living, and they want to be knowledge workers in the jobs in which they contribute to society—that is, their volunteer work. If nonprofit organizations want to attract and hold them, they have to put their competence and knowledge to work. They have to offer meaningful achievement.

Training, Training, and Training

Many nonprofits systematically recruit for such people. Seasoned volunteers are assigned to scan the newcomers—the new member in a church or synagogue, the neighbor who collects for the Red Cross—to find those with leadership talent and persuade them to try themselves in more demanding assignments. Then senior staff (either a full-timer on the payroll or a seasoned volunteer) interviews the newcomers to assess their strengths and place them accordingly. Volunteers may also be assigned both a mentor and a supervisor with whom they work out their performance goals. These advisers are two different people, as a rule, and both, ordinarily, volunteers themselves.

The Girl Scouts, which employs 730,000 volunteers and only 6,000 paid staff for 3.5 million girl members, works this way. A volunteer typically starts by driving youngsters once a week to a meeting. Then a more seasoned volunteer draws her into other work—accompanying Girl Scouts selling cookies door-to-door, assisting a Brownie leader on a camping trip. Out of this step-by-step process evolve the volunteer boards of the local councils and, eventually, the Girl Scouts governing organ, the National Board. Each step, even the very first, has its own compulsory training program, usually conducted by a woman who is herself a volunteer. Each step has specific performance standards and performance goals.

What do these unpaid staff people themselves demand? What makes them stay? And, of course, they can leave at any time. Their first and most important demand is that the nonprofit have a clear mission, one that drives everything the organization does. A senior vice president in a large regional bank has two small children. Yet she just took over as chair of the state chapter of Nature Conservancy, which finds, buys, and manages endangered natural ecologies. "I love my job," she said, when I asked her why she took on such heavy additional work, "and of course the bank has a creed. But it doesn't really know what it contributes. At Nature Conservancy, I know what I am here for."

The second thing this new breed requires, indeed demands, is training, training, and more training. And, in turn, the most effective way to motivate and hold veterans is to recognize their expertise and use them to train newcomers. Then these knowledge workers demand responsibility—above all, for thinking through and setting their own performance goals. They expect to be consulted and to participate in making decisions that affect their work and the work of the organization as a whole. And they expect opportunities for advancement, that is, a chance to take on more demanding assignments and more responsibility as their performance warrants. That is why a good many nonprofits have developed career ladders for their volunteers.

Supporting all this activity is accountability. Many of today's knowledge-worker volunteers insist on having their performance reviewed against preset objectives at least once a year. And increasingly, they expect their organizations to remove nonperformers by moving them to other assignments that better fit their capacities or by counseling them to leave. "It's worse than the Marine Corps boot camp," says the priest in charge of volunteers in the midwestern diocese, "but we have four hundred people on the waiting list." One large and growing midwestern art museum requires of its volunteers—board members, fund-raisers, docents, and the people who edit the museum's newsletter—that they set their goals each year, appraise themselves against those goals, and resign when they fail to meet their goals two years in a row. So does a fair-sized Jewish organization working on college campuses.

These volunteer professionals are still a minority, but a significant one—perhaps one-tenth of the total volunteer population. And they are growing in numbers and, more important, in their impact on the nonprofit sector. Increasingly, nonprofits say what the minister in a large pastoral church says: "There is no laity in this church; there are only pastors, a few paid, most unpaid."

A Warning to Business

This move from nonprofit volunteer to nonpaid professional may be the most important development in American society today. We hear a great deal about the decay and dissolution of family and community and about the loss of values. And, of course, there is reason for concern. But the nonprofits are generating a powerful countercurrent. They are forging new bonds of community, a new commitment to active citizenship, to social responsibility, to values.

And surely what the nonprofit contributes to the volunteer is as important as what the volunteer contributes to the nonprofit. Indeed, it may be fully as important as the service, whether religious, educational, or welfare related, that the nonprofit provides in the community.

This development also carries a clear lesson for business. Managing the knowledge worker for productivity is the next great challenge for American management. The nonprofits are showing us how to do that. It requires a clear mission, careful placement and continual learning and teaching, management by objectives and self-control, high demands but corresponding responsibility, and accountability for performance and results.

There is also, however, a clear warning to American business in this transformation of volunteer work. The students in the program for senior and middle-level executives in which I teach work in a wide diversity of businesses: banks and insurance companies, large retail chains, aerospace and computer companies, real estate developers, and many others. But most of them also serve as volunteers in nonprofits—in a church, on the board of the college they graduated from, as scout leaders, with the YMCA or the Community Chest or the local symphony orchestra. When I ask them why they do it, far too many give the same answer: In my paying job there isn't much challenge, not enough opportunity for achievement, not enough responsibility; and there is no mission, there is only expediency.

5.

SOCIAL IMPACTS
AND SOCIAL PROBLEMS

Social responsibilities—whether of a business, a hospital, or a university—may arise in two areas. They may emerge out of the social impacts of the institution. Or they arise as problems of the society itself. Both are of concern to management because the institution that managers manage lives of necessity in society and community. But otherwise, the two areas are different. The first deals with what an institution does *to* society. The second is concerned with what an institution can do *for* society.

The modern organization exists to provide a specific service to society. It therefore has to be in society. It has to be in a community, has to be a neighbor, has to do its work within a social setting. But also, it has to employ people to do its work. Its *social impacts* inevitably go beyond the specific contribution it exists to make.

The purpose of the hospital is not to employ nurses and cooks. It is patient care. But to accomplish this purpose, nurses and cooks are needed. And in no time at all they form a work community with its own community tasks and community problems.

The purpose of a ferroalloy plant is not to make noise or to release noxious fumes. It is to make high-performance metals that serve the customer. But in order to do this, it produces noise, creates heat, and releases fumes.

These impacts are incidental to the purpose of the organization. But in large measure they are inescapable by-products.

Social problems, by contrast, are dysfunctions of society rather than impacts of the organization and its activities.

Since the institution can exist only within the social environment, is indeed an organ of society, such social problems affect the institution. They are of concern to it even if the community itself sees no problem and resists any attempt to tackle it.

A healthy business, a healthy university, a healthy hospital cannot exist in a sick society. Management has a self-interest in a healthy society, even though the cause of society's sickness is none of management's making.

Responsibility for Impacts

One is responsible for one's impacts, whether they are intended or not. This is the first rule. There is no doubt regarding management's responsibility for the social impacts of its organization. They are management's business.

It is not enough to say, But the public doesn't object. It is, above all, not enough to say that any action to come to grips with a particular problem is going to be "unpopular," is going to be "resented" by one's colleagues and one's associates, and is not required. Sooner or later society will come to regard any such impact as an attack on its integrity and will exact a high price from those who have not responsibly worked on eliminating the impact or on finding a solution to the problem.

Here is one example.

In the late 1940s and early '50s, one American automobile company tried to make the American public safety-conscious. Ford introduced cars with seat belts. But sales dropped catastrophically. The company had to withdraw the cars with seat belts and abandoned the whole idea. When, fifteen years later, the American driving public became safety-conscious, the car manufacturers were sharply attacked for their "total lack of concern with safety" and for being "merchants of death." And the resulting regulations were written as much to punish the companies as to protect the public.

The first job of management is, therefore, to identify and to anticipate impacts—coldly and realistically. The question is not, Is what we do right? It is, Is what we do what society and the customer pay us for?

How to Deal with Impacts

Identifying incidental impacts of an institution is the first step. But how does management deal with them? The objective is clear: impacts on society and economy, community, and individual that are not in themselves the purpose and mission of the institution should be kept to a minimum and should preferably be eliminated altogether. The fewer such impacts the better, whether the impact is within the institution, on the social environment, or on the physical environment.

Wherever an impact can be eliminated by dropping the activity that causes it, that is therefore the best—indeed the only truly good—solution.

In most cases the activity cannot, however, be eliminated. Hence, there is need for systematic work at eliminating the impact— or at least at minimizing it—while maintaining the underlying activity itself. The ideal approach is to make the elimination of impacts into a profitable business opportunity. One example is the way Dow Chemical, one of the leading U.S. chemical companies, has for almost twenty years tackled air and water pollution. Dow decided, shortly after World War II, that air and water pollution was an undesirable impact that had to be eliminated. Long before the public outcry about the environment, Dow adopted a zero-pollution policy for its plants. It then set about systematically to develop the polluting substances it removes from smokestack gases and watery effluents into salable products and to create uses and markets for them.

Another example is the Du Pont Industrial Toxicity Laboratory. Du Pont, in the 1920s, became aware of the toxic side effects of many of its industrial products and set up a laboratory to test for toxicity and to develop processes to eliminate the poisons. Du Pont started out to eliminate an impact that at that time every other chemical manufacturer took for granted. But then Du Pont decided to develop toxicity control of industrial products into a separate business. The Industrial Toxicity Laboratory works not only for Du Pont but for a wide variety of customers for whom it develops non-poisonous compounds, whose products it tests for toxicity, and so on. Again, an impact has been eliminated by making it into a business opportunity.

To make elimination of an impact into a business opportunity should always be attempted. But it cannot be done in many cases. More often eliminating an impact means increasing the costs. What was an "externality" for which the general public paid becomes business cost. It therefore becomes a competitive disadvantage unless everybody in the industry accepts the same rule. And this, in most cases, can be done only by regulation—that means by some form of public action.

Whenever an impact cannot be eliminated without an increase in cost, it becomes incumbent upon management to think ahead and work out the regulation that is most likely to solve the problem at the minimum cost and with the greatest benefit to public and business alike. And it is then management's job to work at getting the right regulation enacted.

Management—and not only business management—has shunned this responsibility. The traditional attitude has always been that "no regulation is the best regulation." But this applies only when an impact can be made into a business opportunity. Where elimination of an impact requires a restriction, regulation is in the interest of business, and especially in the interest of responsible business. Otherwise, it will be penalized as "irresponsible," while the unscrupulous, the greedy, and the stupid cash in.

And to expect that there will be no regulation is willful blindness.

The fact that the public today sees no issue is not relevant. Indeed, it is not even relevant that the public today—as it did in the Ford example above—resists actively any attempts on the part of farsighted business leaders to prevent a crisis. In the end, there is the scandal.

Any solution to an impact problem requires trade-offs. Beyond a certain level elimination of an impact costs more in money or in energy, in resources or in lives, than the attainable benefit. A decision has to be made on the optimal balance between costs and benefits. This is something people in an industry understand, as a rule. But no one outside does—and so the outsider's solution tends to ignore the trade-off problem altogether.

Responsibility for social impacts is a management responsibility—not because it is a social responsibility, but because it is a business responsibility. The ideal is to make elimination of such an impact into a business opportunity. But wherever that cannot be done, the design of the appropriate regulation with the optimal trade-off balance—and public discussion of the problem and promotion of the best regulatory solution—is management's job.

Social Problems as Business Opportunities

Social problems are dysfunctions of society and—at least potentially—degenerative diseases of the body politic. They are ills. But for the management of institutions, and, above all, for business management, they represent challenges. They are major sources of opportunity. For it is the function of business—and to a lesser degree of the other main institutions—to satisfy a social need and at the same time serve themselves by making resolution of a social problem into a business opportunity.

It is the job of business to convert change into innovation, that is, into new business. And it is a poor businessman who thinks that innovation refers to technology alone. Social change and social innovation have throughout business history been at least as important as technology. After all, the major industries of the nineteenth century were, to a very large extent, the result of converting the new social environment—the industrial city—into a business opportunity and into a business market. This underlay the rise of lighting, first by gas and then by electricity, of the streetcar and the interurban trolley, of telephone, newspaper, and department store—to name only a few.

The most significant opportunities for converting social problems into business opportunities may therefore not lie in new technologies, new products, and new services. They may lie in *solving* the social problem, that is, in social innovation, which then directly and indirectly benefits and strengthens the company or the industry.

The experience of some of the most successful businesses is largely the result of such social innovation.

The years immediately prior to World War I were years of great labor unrest in the United States, growing labor bitterness, and high unemployment. Hourly wages for skilled men ran as low as fifteen cents in many cases. It was against this background that the Ford Motor Company, in the closing days of 1913, announced that it would pay a guaranteed five-dollar-a-day wage to every one of its workers—two to three times what was then standard. James Couzens, the company's general manager, who had forced this decision on his reluctant partner, Henry Ford, knew perfectly well that his company's wage bill would almost triple overnight. But he became convinced that the workmen's sufferings were so great that only radical and highly visible action could have an effect. Couzens also expected that Ford's actual labor cost, despite the tripling of the wage rate, would go down—and events soon proved him right. Before Ford changed the whole labor economy of the United States with one announcement, labor turnover at the Ford Motor Company had been so high that, in 1912, sixty thousand men had to be hired to retain ten thousand workers. With the new wage, turnover almost disappeared. The resulting savings were so great that despite sharply rising costs for all materials in the next few years, Ford could produce and sell its Model T at a lower price and yet make a larger profit per car. It was the saving in labor cost produced by a drastically higher wage that gave Ford market domination. At the same time, Ford's action transformed American industrial society. It established the American workingman as fundamentally middle class.

Social problems that management action converts into opportunities soon cease to be problems. The others, however, are likely to become "chronic complaints," if not "degenerative diseases."

Not every social problem can be resolved by making it into an opportunity for contribution and performance. Indeed, the most serious of such problems tend to defy this approach.

What then is the social responsibility of management for these social problems that become chronic complaints or degenerative diseases?

They are management's problems. The health of the enterprise is management's responsibility. A healthy business and a sick society are hardly compatible. Healthy businesses require a healthy, or at least a functioning, society. The health of the community is a prerequisite for successful and growing business.

And it is foolish to hope that these problems will disappear if only one looks the other way. Problems go away because someone does something about them.

To what extent should business—or any other of the special-purpose institutions of our society—be expected to tackle such a problem that did not arise out of an impact of theirs and that cannot be converted into an opportunity for performance of the institution's purpose and mission? To what extent should these institutions, business, university, or hospital, even be permitted to take responsibility?

Today's rhetoric tends to ignore that question. "Here is," John Lindsay, former mayor of New York, said, "the black ghetto. No one knows what to do with it. Whatever government, social workers, or community action try, things seem only to get worse. *Therefore,* big business better take responsibility."

That Mayor Lindsay frantically looks for someone to take over is understandable; and the problem that is defeating him is indeed desperate and a major threat to his city, to American society, and to the Western world altogether. But is it enough to make the problem of the black ghetto the social responsibility of management? Or are there limits to social responsibility? And what are they?

The Limits of Social Responsibility

The manager is a servant. His or her master is the institution being managed and the first responsibility must therefore be to it. The manager's first task is to make the institution, whether business, hospital, school, or university, perform the function and make the contribution for the sake of which it exists. The manager who uses a position at the head of a major institution to become a public figure and to take leadership with respect to social problems, while the company or the university erodes through neglect, is not a statesman, but is irresponsible and false to his trust.

The institution's performance of its specific mission is also society's first need and interest. Society does not stand to gain but to lose if the performance capacity of the institution in its own specific task is diminished or impaired. Performance of its function is the institution's first social responsibility. Unless it discharges its performance responsibly, it cannot discharge anything else. A bankrupt business is not a desirable employer and is unlikely to be a good neighbor in a community. Nor will it create the capital for tomorrow's jobs and the opportunities for tomorrow's workers. A university that fails to prepare tomorrow's leaders and professionals is not socially responsible, no matter how many "good works" it engages in.

Above all, management needs to know the *minimum profitability* required by the risks of the business and by its commitments to the future. It needs this knowledge for its own decisions. But it needs it just as much to explain its decisions to others—the politicians, the press, the public. As long as managements remain the prisoners of their own ignorance of the objective need for, and function of, profit—i.e., as long as they think and argue in terms of the "profit motive"—they will be able neither to make rational decisions with respect to social responsibilities, nor to explain these decisions to others inside and outside the business.

Whenever a business has disregarded the limitation of economic performance and has assumed social responsibilities that it could not support economically, it has soon gotten into trouble.

The same limitation on social responsibility applies to noneconomic institutions. There, too, the manager is duty-bound to preserve the performance capacity of the institution in his care. To jeopardize it, no matter how noble the motive, is irresponsibility. These institutions, too, are capital assets of society on the performance of which society depends.

This, to be sure, is a very unpopular position to take. It is much more popular to be "progressive." But managers, and especially managers of key institutions of society, are not being paid to be heroes to the popular press. They are being paid for performance and responsibility.

To take on tasks for which one lacks competence is irresponsible behavior. It is also cruel. It raises expectations that will then be disappointed.

An institution, and especially a business enterprise, has to acquire whatever competence is needed to take responsibility for its impacts. But in areas of social responsibility other than impacts, right and duty to act are limited by competence.

In particular an institution better refrain from tackling tasks that do not fit into its value system. Skills and knowledge are fairly easily acquired. But one cannot easily change personality. No one is likely to do well in areas that he does not respect. If a business or any other institution tackles such an area because there is a social need, it is unlikely to put its good people on the task and to support them adequately. It is unlikely to understand what the task involves. It is almost certain to do the wrong things. As a result, it will do damage rather than good.

Management therefore needs to know at the very least what it and its institution are truly *incompetent* for. Business, as a rule, will be in this position of absolute incompetence in an "intangible" area. The strength of business is accountability and measurability. It is the discipline of market test, productivity measurements, and profitability requirement. Where these are lacking, businesses are essentially out of their depth. They are also out of fundamental sympathy, that is, outside their own value systems. Where the criteria of performance are intangible, such as "political" opinions and emotions, community approval or disapproval, mobilization of community energies and structuring of power relations, business is unlikely to feel comfortable. It is unlikely to have respect for the values that matter. It is, therefore, most unlikely to have competence.

In such areas it is, however, often possible to define goals clearly and measurably for *specific partial tasks*. It is often possible to convert parts of a problem that by itself lies outside the competence of business into work that fits the competence and value system of the business enterprise.

No one in America has done very well in training hard-core unemployable black teenagers for work and jobs. But business has done far less badly than any other institution: schools, government programs, community agencies. This task can be identified. It can be defined. Goals can be set. And performance can be measured. And then business can perform.

The Limits of Authority

The most important limitation on social responsibility is the limitation of authority. The constitutional lawyer knows that there is no such word as "responsibility" in the political dictionary. The appropriate term is "responsibility *and* authority." Whoever claims authority thereby assumes responsibility. But whoever assumes responsibility thereby claims authority. The two are but different sides of the same coin. To assume social responsibility therefore always means to claim authority.

Again, the question of authority as a limit on social responsibility does not arise in connection with the impacts of an institution. For the impact is the result of an exercise of authority, even though purely incidental and unintended. And then responsibility follows.

But where business or any other institution of our society of organizations is asked to assume social responsibility for one of the problems or ills of society and community, management needs to think through whether the authority implied in the responsibility is legitimate. Otherwise it is usurpation and irresponsible.

Every time the demand is made that business take responsibility for this or that, one should ask, Does business have the authority and should it have it? If business does not have and should not have authority—and in a great many areas it should not have it—then responsibility on the part of business should be treated with grave suspicion. It is not responsibility; it is lust for power.

Ralph Nader, the American consumerist, sincerely considers himself a foe of big business and is accepted as such by business and by the general public. Insofar as Nader demands that business take responsibility for product quality and product safety, he is surely concerned with legitimate business responsibility, i.e., with responsibility for performance and contribution.

But Ralph Nader demands, above all, that big business assume responsibility in a multitude of areas beyond products and services. This, if acceded to, can lead only to the emergence of the managements of the big corporations as the ultimate power in a vast number of areas that are properly some other institution's field.

And this is, indeed, the position to which Nader—and other advocates of unlimited social responsibility—are moving rapidly. One of the Nader task forces published a critique of the Du Pont Company and its role in the small state of Delaware, where Du Pont has its headquarters and is a major employer. The report did not even discuss economic performance; it dismissed as irrelevant that Du Pont, in a period of general inflation, consistently lowered the prices for its products, which are, in many cases, basic materials for the American economy. Instead it sharply criticized Du Pont for not using its economic power to force the citizens of the state to attack a number of social problems, from racial discrimination to health care to public schools. Du Pont, for not taking responsibility for Delaware society, Delaware politics, and Delaware law, was called grossly remiss in its social responsibility.

One of the ironies of this story is that the traditional liberal or left-wing criticism of the Du Pont Company for many years has been the exact opposite, i.e., that Du Pont, by its very prominence in a small state, "interferes in and dominates" Delaware and exercises "illegitimate authority."

Management must resist responsibility for a social problem that would compromise or impair the performance capacity of its business (or its university or its hospital). It must resist when the demand goes beyond the institution's competence. It must resist when responsibility would, in fact, be illegitimate authority. But then, if the problem is a real one, it better think through and offer an alternative approach. If the problem is serious, something will ultimately have to be done about it.

Managements of all major institutions, including business enterprise, need, too, to concern themselves with serious ills of society. If at all possible, they should convert the solution of these problems into an opportunity for performance and contribution. At the least they should think through what the problem is and how it might be tackled. They cannot escape concern; for this society of organizations has no one else to be concerned about real problems. In this society, managers of institutions are the leadership group.

But we also know that a developed society needs performing institutions with their own autonomous management. It cannot function as a totalitarian society. Indeed, what characterizes a developed society—and indeed makes it a developed one—is that most of its social tasks are carried out in and through organized institutions, each with its own autonomous management. These organizations, including most of the agencies of our government, are special-purpose institutions. They are organs of our society for specific performance in a specific area. The greatest contribution they can make, their greatest social responsibility, is performance of their function. The greatest social irresponsibility is to impair the performance capacity of these institutions by tackling tasks beyond their competence or by usurpation of authority in the name of social responsibility.

The Ethics of Responsibility

Countless sermons have been preached and printed on the ethics of business or the ethics of the businessman. Most have nothing to do with business and little to do with ethics.

One main topic is plain, everyday honesty. Businessmen, we are told solemnly, should not cheat, steal, lie, bribe, or take bribes. But nor should anyone else. Men and women do not acquire exemption from ordinary rules of personal behavior because of their work or job. Nor, however, do they cease to be human beings when appointed vice president, city manager, or college dean. And there has always been a number of people who cheat, steal, lie, bribe, or take bribes. The problem is one of moral values and moral education, of the individual, of the family, of the school. But neither is there a separate ethics of business, nor is one needed.

All that is needed is to mete out stiff punishments to those—whether business executives or others—who yield to temptation.

The other common theme in the discussion of ethics in business has nothing to do with ethics. It would indeed be nice to have fastidious leaders. Alas, fastidiousness has never been prevalent among leadership groups, whether kings and counts, priests, or generals, or even "intellectuals" such as the painters and humanists of the Renaissance, or the "literati" of the Chinese tradition. All a fastidious man can do is withdraw personally from activities that violate his self-respect and his sense of taste.

Lately these old sermon topics have been joined, especially in the United States, by a third one: managers, we are being told, have an "ethical responsibility" to take an active and constructive role in their community, to serve community causes, give of their time to community activities, and so on.

Such activities should, however, never be forced on them, nor should managers be appraised, rewarded, or promoted according to their participation in voluntary activities. Ordering or pressuring managers into such work is abuse of organizational power and illegitimate.

But, while desirable, community participation of managers has nothing to do with ethics, and not much to do with responsibility. It is the contribution of an individual in his or her capacity as a neighbor and citizen. And it is something that lies outside the manager's job and outside managerial responsibility.

A problem of ethics that is peculiar to the manager arises from the managers of institutions being *collectively* the leadership groups of the society of organizations. But *individually* a manager is just another fellow employee.

It is therefore inappropriate to speak of managers as leaders. They are "members of the leadership group." The group, however, does occupy a position of visibility, of prominence, and of authority. It therefore has responsibility.

But what are the responsibilities, what are the ethics of the individual manager, as a member of the leadership group?

Essentially being a member of a leadership group is what traditionally has been meant by the term "professional." Membership in such a group confers status, position, prominence, and authority. It also confers duties. To expect every manager to be a leader is futile. There are, in a developed society, thousands, if not millions, of managers—and leadership is always the rare exception and confined to a very few individuals. But as a member of a leadership group a manager stands under the demands of professional ethics—the demands of an ethic of responsibility.

Not Knowingly to Do Harm

The first responsibility of a professional was spelled out clearly, twenty-five hundred years ago, in the Hipprocratic oath of the Greek physician: *Primum non nocere*—"Above all, not knowingly to do harm."

No professional, be he doctor, lawyer, or manager, can promise that he will indeed do good for his client. All he can do is try. But he can promise that he will not knowingly do harm. And the client, in turn, must be able to trust the professional not knowingly to do him harm. Otherwise he cannot trust him at all. The professional has to have autonomy. He cannot be controlled, supervised, or directed by the client. He has to be private in that his knowledge and his judgment have to be entrusted with the decision. But it is the foundation of his autonomy, and indeed its rationale, that he see himself as "affected with the public interest." A professional, in other words, is private in the sense that he is autonomous and not subject to political or ideological control. But he is public in the sense that the welfare of his client sets limits to his deeds and words. And *Primum non nocere,* "not knowingly to do harm," is the basic rule of professional ethics, the basic rule of an ethics of public responsibility.

The manager who fails to think through and work for the appropriate solution to an impact of his business because it makes him "unpopular in the club" knowingly does harm. He knowingly abets a cancerous growth. That this is stupid has been said. That this always in the end hurts the business or the industry more than a little temporary "unpleasantness" would have hurt has been said, too. But it is also gross violation of professional ethics.

But there are other aspects to this issue as well. American managers, in particular, tend to violate the rule usually without knowing that they do so, and in so doing they cause harm, especially with respect to:

- Executive compensation
- Use of benefit plans to impose "golden fetters" on people in the company's employ
- Their profit rhetoric

Their actions and their words in these areas tend to cause social disruption. They tend to conceal reality and to create disease, or at least social hypochondria. They tend to misdirect and to prevent understanding. And this is grievous social harm.

The fact of increasing income equality in U.S. society is quite clear. Yet the popular impression is one of rapidly increasing inequality. This is illusion; but it is a dangerous illusion. It corrodes. It destroys mutual trust between groups that have to live together and work together. It can only lead to political measures that, while doing no one any good, can seriously harm society, economy, and the manager as well.

The $500,000 a year that the chief executive of one of the giant corporations is being paid is largely "make-believe money." Its function is status rather than income. Most of it, whatever tax loopholes the lawyers might find, is immediately taxed away. And the "extras" are simply attempts to put a part of the executive's income into a somewhat lower tax bracket. Economically, in other words, neither serves much purpose. But socially and psychologically they "knowingly do harm." They cannot be defended.

What is pernicious, however, is the illusion of inequality. The basic cause is the tax laws. But the managers' willingness to accept, and indeed to play along with, an antisocial tax structure is a major contributory cause. And unless managers realize that this violates the rule "not knowingly to do damage," they will, in the end, be the main sufferers.

A second area in which the manager of today does not live up to the commitment of *Primum non nocere* is closely connected with compensation.

Retirement benefits, extra compensation, bonuses, and stock options are all forms of compensation. From the point of view of the enterprise—but also from the point of view of the economy—these are "labor costs" no matter how they are labeled. They are treated as such by managements when they sit down to negotiate with the labor union. But increasingly, if only because of the bias of the tax laws, these benefits are being used to tie an employee to his employer. They are being made dependent on staying with the same employer, often for many years. And they are structured in such a way that leaving a company's employ entails drastic penalties and actual loss of benefits that have already been earned and that, in effect, constitute wages relating to the past employment.

Golden fetters do not strengthen the company. People who know that they are not performing in their present employment—that is, people who are clearly in the wrong place—will often not move but stay where they know they do not properly belong. But if they stay because the penalty for leaving is too great, they resist and resent it. They know that they have been bribed and were too weak to say no. They are likely to be sullen, resentful, and bitter the rest of their working lives.

Pension rights, performance bonuses, participation in profits, and so on, have been "earned" and should be available to the employee without restricting his rights as a citizen, an individual, and a person. And, again, managers will have to work to get the tax law changes that are needed.

Managers, finally, through their rhetoric, make it impossible for the public to understand economic reality. This violates the requirement that managers, being leaders, not knowingly do harm. This is particularly true of the United States but also of Western Europe. For in the West, managers still talk constantly of the profit motive. And they still define the goal of their business as profit maximization. They do not stress the objective function of profit. They do not talk of risks—or very rarely. They do not stress the need for capital. They almost never even mention the cost of capital, let alone that a business has to produce enough profit to obtain the capital it needs at minimum cost.

Managers constantly complain about the hostility to profit. They rarely realize that their own rhetoric is one of the main reasons for this hostility. For indeed in the terms management uses when it talks to the public, there is no possible justification for profit, no explanation for its existence, no function it performs. There is only the profit motive, that is, the desire of some anonymous capitalists—and why that desire should be indulged in by society any more than bigamy, for instance, is never explained. But profitability is a crucial *need* of economy and society.

Primum non nocere may seem tame compared with the rousing calls for "statesmanship" that abound in today's manifestos on social responsibility. But, as the physicians found out long ago, it is not an easy rule to live up to. Its very modesty and self-constraint make it the right rule for the ethics that managers need, the ethics of responsibility.

6.

MANAGEMENT'S NEW
PARADIGMS

Basic assumptions about reality are the paradigms of a social science, such as management. They are usually held subconsciously by the scholars, the writers, the teachers, the practitioners in the field, and are incorporated into the discipline by their various formulations. Thus those assumptions by this select group of people largely determine what the discipline assumes to be reality.

The discipline's basic assumptions about reality determine what it focuses on. They determine what a discipline considers "facts," and indeed what the discipline considers itself to be all about. The assumptions also largely determine what is being disregarded or is being pushed aside as an "annoying exception."

Yet, despite their importance, the assumptions are rarely analyzed, rarely studied, rarely challenged—indeed rarely even made explicit.

For a social discipline such as management, the assumptions are actually a good deal more important than are the paradigms for a natural science. The paradigm—that is, the prevailing general theory—has no impact on the natural universe. Whether the paradigm states that the sun rotates around the earth or that, on the contrary, the earth rotates around the sun has no effect on sun and earth. A natural science deals with the behavior of <u>objects</u>. But a social discipline such as management deals with the behavior of <u>people</u> and <u>human institutions</u>. Practitioners will therefore tend to act and to behave as the discipline's assumptions tell them to. Even more important, the reality of a natural science, the physical universe and its laws, do not change (or if they do only over eons rather than over centuries, let alone over decades). The social universe has no "natural laws" of this kind. It is thus subject to continual change. And this means that assumptions that were valid yesterday can become invalid and, indeed, totally misleading in no time at all.

What matters most in a social discipline such as management are therefore the basic assumptions. And a <u>change</u> in the basic assumptions matters even more.

Since the study of management first began—and it truly did not emerge until the 1930s—<u>two sets</u> of assumptions regarding the <u>realities</u> of management have been held by most scholars, most writers and most practitioners:

One set of assumptions underlies the <u>discipline</u> of management:

1. Management is <u>business</u> management.
2. There is—or there must be—<u>one</u> right *organization structure*.
3. There is—or there must be—<u>one</u> right way to *manage people*.

Another set of assumptions underlies the <u>practice</u> of management:

1. Technologies, markets and end uses are *given*.
2. Management's scope is *legally* defined.
3. Management is internally focused.
4. The economy as defined by national boundaries is the "ecology" of enterprise and management.

Management Is Business Management

For most people, inside and outside management, this assumption is taken as self-evident. Indeed management writers, management practitioners, and the laity do not even hear the word "management"; they automatically hear <u>business management</u>.

This assumption regarding the universe of management is of fairly recent origin. Before the 1930s the few writers and thinkers who concerned themselves with management—beginning with Frederick Winslow Taylor around the turn of the century and ending with Chester Barnard just before World War II—all assumed that business management is just a subspecies of general management and basically no more different from the management of any other organization than one breed of dogs is from another.

What led to the identification of management with business management was the Great Depression with its hostility to business and its contempt for business executives. In order not to be tarred with the business brush, management in the public sector was rechristened "public administration" and proclaimed a separate discipline—with its own university departments, its own terminology, its own career ladder. At the same time—and for the same reason—what had begun as a study of management in the rapidly growing hospital (e.g., by Raymond Sloan, the younger brother of GM's Alfred Sloan) was split off as a separate discipline and christened "hospital administration."

Not to be called "management" was, in other words, "political correctness" in the Depression years.

In the postwar period, however, the fashion turned. By 1950 "business" had become a "good word"—largely the result of the performance during World War II of American *business* management. And then very soon business management became "politically correct" as a field of study, above all. And ever since, management has remained identified in the public mind as well as in academia with business management.

Now, we are beginning to unmake this sixty-year-old mistake—as witness the renaming of so many "business schools" as "schools of management," the rapidly growing offerings in "nonprofit management" by these schools, the emergence of "executive management programs" recruiting both business and nonbusiness executives, or the emergence of departments of "pastoral management" in divinity schools.

But the assumption that management is business management still persists. It is therefore important to assert—and to do so loudly—that management is <u>not</u> business management—any more than, say, medicine is obstetrics.

There are, of course, differences in management among different organizations—mission defines strategy, after all, and strategy defines structure. There surely are differences in managing a chain of retail stores and managing a Catholic diocese (though amazingly fewer than either chain stores or bishops might believe); in managing an air base, a hospital, and a software company. But the greatest differences are in the terms individual organizations use. Otherwise the differences are mainly in application rather than in principles. There are not even tremendous differences in tasks and challenges.

The first conclusion of this analysis of the <u>assumptions</u> that must underlie management to make productive both its study and its practice is therefore:

Management is the specific and distinguishing organ of any and all organizations.

The One Right Organization

Concern with management and its study began with the sudden emergence of large organizations—business, governmental civil service, the large standing army—which was the novelty of late-nineteenth-century society.

And from the very beginning more than a century ago, the study of organization has rested on one assumption:

There is—or there must be—one right organization.

What is presented as the "one right organization" has changed more than once. But the search for the one right organization has continued and continues today.

It was World War I that made clear the need for a formal organization structure. But it was also World War I that showed that Fayol's (and Carnegie's) functional structure was not the one right organization. Immediately after World War I first Pierre S. Du Pont (1870–1954) and then Alfred Sloan (1875–1966) developed the principle of *decentralization*. And now, in the last few years, we have come to tout the *team* as the one right organization for pretty much everything.

By now, however, it should have become clear that there is no such thing as the one right organization. There are only organizations, each of which has distinct strengths, distinct limitations, and specific applications. It has become clear that organization is not an absolute. It is a *tool* for making people productive in working together. As such, a given organization structure fits certain tasks in certain conditions and at certain times.

One hears a great deal today about "the end of hierarchy." This is blatant nonsense. In any institution there has to be a final authority, that is, a "boss"—someone who can make the final decisions and who can expect them to be obeyed. In a situation of common peril—and every institution is likely to encounter it sooner or later—survival of all depends on clear command. If the ship goes down, the captain does not call a meeting, the captain gives an order. And if the ship is to be saved, everyone must obey the order, must know exactly where to go and what to do, and do it without "participation" or argument. "Hierarchy," and the unquestioning acceptance of it by everyone in the organization, is the only hope in a crisis.

Other situations within the same institution require deliberation. Others still require teamwork—and so on.

Organization theory assumes that institutions are homogeneous and that, therefore, the entire enterprise should be organized the same way.

But in any one enterprise—probably even in Fayol's "typical manufacturing company"—there is need for a number of different organization structures coexisting side by side.

Managing foreign currency exposure is an increasingly critical—and increasingly difficult—task in a world economy. It requires total centralization. No one unit of the enterprise can be permitted to handle its own foreign currency exposures. But in the same enterprise servicing the customer, especially in high-tech areas, requires almost complete local autonomy—going way beyond traditional decentralization. Each of the individual service people has to be the "boss," with the rest of the organization taking its direction from them.

Certain forms of research require a strict functional organization with all specialists "playing their instrument" by themselves. Other kinds of research, however, especially research that involves decision-making at an early stage (e.g., some pharmaceutical research), require teamwork from the beginning. And the two kinds of research often occur side by side and in the same research organization.

The belief that there must be one right organization is closely tied to the fallacy that management is business management. If earlier students of management had not been blinkered by this fallacy but had looked at nonbusinesses, they would soon have found that there are vast differences in organization structure according to the nature of the task.

A Catholic diocese is organized very differently from an opera. A modern army is organized very differently from a hospital.

There are indeed some "principles" of organization.

One is surely that organization has to be transparent. People have to know and have to understand the organization structure they are supposed to work in. This sounds obvious—but it is far too often violated in most institutions (even in the military).

Another principle I have already mentioned: Someone in the organization must have the authority to make the final decision in a given area. And someone must clearly be in command in a <u>crisis</u>. It also is a sound principle that authority be commensurate with responsibility.

It is a sound principle that one person in an organization should have only one "master." There is wisdom to the old proverb of the Roman law that a slave who has three masters is a free man. It is a very old principle of human relations that no one should be put into a conflict of loyalties—and having more than one "master" creates such a conflict (which, by the way, is the reason that the "jazz combo" team, so popular now, is so difficult—every one of its members has two masters, the head of the specialty function, for example, engineering, and the team leader). It is a sound, structural principle to have the fewest layers, that is, to have an organization that is as "flat" as possible—if only because, as information theory tells us, "every relay doubles the noise and cuts the message in half."

But these principles do not tell us *what to do*. They only tell us what not to do. They do not tell us what will work. They tell us what is unlikely to work. These principles are not too different from the ones that inform an architect's work. They do not tell him what kind of building to build. They tell him what the restraints are. And this is pretty much what the various principles of organization structure do.

One implication: *Individuals* will have to be able to work at one and the same time in different organization structures. For one task they will work in a team. But for another task they will have to work—and at the same time—in a command-and-control structure. The same individual who is a "boss" within his or her own organization is a "partner" in an alliance, a minority participation, a joint venture, and so on. Organizations, in other words, will have to become part of the executive's toolbox.

Even more important: We need to go to work on studying the strengths and the limitations of different organizations. For what tasks are what organizations most suitable? For what tasks are what organizations least suitable? And when, in the performance of a task, should we switch from one kind of organization to another?

One area in which research and study are particularly needed is the <u>organization of top management</u>.

And I doubt that anyone would assert that we really know how to organize the top management job, whether in a business, a university, a hospital, or even a modern church.

One clear sign is the growing disparity between our rhetoric and our practice. We talk incessantly about "teams"—and every study comes to the conclusion that the top management job does indeed require a team. Yet we now *practice*—and not only in American industry—the most extreme "personality cult" of CEO supermen. And no one seems to pay the slightest attention in our present worship of these larger-than-life CEOs to the question of how and by what process they are to be succeeded—and yet, succession has always been the ultimate test of any top management and the ultimate test of any institution.

There is, in other words, an enormous amount of work to be done in organizational theory and organization practice—even though both are the oldest areas of organized work and organized practice in management.

The pioneers of management a century ago were right. *Organizational structure is needed.* The modern enterprise—whether business, civil service, university, hospital, large church, or large military—needs organization just as any biological organization beyond the amoeba needs structure. But the pioneers were wrong in their assumption that there is—or should be—one right organization. Just as there are a great number of different structures for biological organizations, so there are a number of organizations for the social organism that is the modern institution.

Instead of searching for the right organization, management needs to learn to look for, to develop, to test

The organization that fits the task.

The One Right Way to Manage People

In no other area are the basic traditional assumptions held as firmly—though mostly subconsciously—as in respect to people and their management. And in no other area are they so totally at odds with reality and so totally counterproductive.

There is one right way to manage people—or at least there should be.

This assumption underlies practically every book or paper on the management of people. Its most quoted exposition is Douglas McGregor's book *The Human Side of Enterprise* (1960), which asserted that managements have to choose between two and only two different ways of managing people, "Theory X" and "Theory Y," and which then asserted that Theory Y is the only sound one. (A little earlier I had said pretty much the same thing in my 1954 book *The Practice of Management*.) A few years later Abraham H. Maslow (1908–1970) showed in his *Eupsychian Management* (1962; new edition 1995 entitled *Maslow on Management*) that both McGregor and I were dead wrong. He showed conclusively that different people have to be managed differently.

I became an immediate convert—Maslow's evidence is overwhelming. But to date very few people have paid much attention.

On this fundamental assumption that there is—or at least should be—one and only one right way to manage people, rest all the other assumptions about people in organizations and their management.

One of these assumptions is that the people who work for an organization are *employees* of the organization, working full-time, and dependent on the organization for their livelihood and their careers. Another such assumption is that the people who work for an organization are *subordinates*. Indeed, it is assumed that the great majority of these people have either no skill or low skills and do what they are being assigned to do.

Eighty years ago, when these assumptions were first formulated, during and at the end of World War I, they conformed close enough to reality to be considered valid. Today every one of them has become untenable. The majority of people who work for an organization may still be employees of the organization. But a very large and steadily growing minority—though working *for* the organization—are no longer its employees, let alone its full-time employees. They work for an outsourcing contractor, for example, the outsourcing firm that provides maintenance in a hospital or a manufacturing plant, or the outsourcing firm that runs the data-processing system for a government agency or a business. They are "temps" or part-timers. Increasingly they are individual contractors working on a retainer or for a specific contractual period; this is particularly true of the most knowledgeable and therefore the most valuable people working for the organization.

Even if employed full-time by the organization, fewer and fewer people are "subordinates"—even in fairly low-level jobs. Increasingly they are "knowledge workers." And knowledge workers are not subordinates; they are "associates." For, once beyond the apprentice stage, knowledge workers must know more about their job than their boss does—or else they are no good at all. In fact, that they know more about their job than anybody else in the organization is part of the definition of knowledge workers.

Add to this that today's "superiors" usually have not held the jobs their "subordinates" hold—as they did only a few short decades ago and as still is widely assumed they do.

A regimental commander in the army, only a few decades ago, had held every one of the jobs of his subordinates—battalion commander, company commander, platoon commander. The only difference in these respective jobs between the lowly platoon commander and the lordly regimental commander was in the number of people each commands; the work they did was exactly alike. To be sure, today's regimental commanders have commanded troops earlier in their careers—but often for a short period only. They also have advanced through captain and major. But for most of their careers they have held very different assignments—in staff jobs, in research jobs, in teaching jobs, attached to an embassy abroad and so on. They simply can no longer assume that they know what their "subordinate," the captain in charge of a company, is doing or trying to do—they have been captains, of course, but they may have never commanded a company.

Similarly, the vice president of marketing may have come up the sales route. He or she knows a great deal about selling, but knows nothing about market research, pricing, packaging, service, sales forecasting. The marketing vice president therefore cannot possibly tell the experts in the marketing department what they should be doing, and how. Yet they are supposed to be the marketing vice president's "subordinates"—and the marketing vice president is definitely responsible for their performance and for their contribution to the company's marketing efforts.

The same is true for the hospital administrator or the hospital's medical director in respect to the trained knowledge workers in the clinical laboratory or in physical therapy.

To be sure, these associates are "subordinates" in that they depend on the "boss" when it comes to being hired or fired, promoted, appraised and so on. But in his or her own job the superior can perform only if these so-called subordinates take responsibility for *educating* him or her, that is, for making the "superior" understand what market research or physical therapy can do and should be doing, and what "results" are in their respective areas. In turn, these "subordinates" depend on the superior for direction. They depend on the superior to tell them what the "score" is.

Their relationship, in other words, is far more like that between the conductor of an orchestra and the instrumentalist than it is like the traditional superior/subordinate relationship. The superior in an organization employing knowledge workers cannot, as a rule, do the work of the supposed subordinate any more than the conductor of an orchestra can play the tuba. In turn, the knowledge worker is dependent on the superior to give direction and, above all, to define what the "score" is for the entire organization, that is, what are its standards and values, performance and results. And just as an orchestra can sabotage even the ablest conductor—and certainly even the most autocratic one—a knowledge organization can easily sabotage even the ablest, let alone the most autocratic, superior.

Altogether, an increasing number of people who are full-time employees have to be managed as if they were *volunteers*. They are paid, to be sure. But knowledge workers have mobility. They can leave. They own their "means of production," which is their knowledge.

We have known for fifty years that money alone does not motivate to perform. Dissatisfaction with money grossly demotivates. Satisfaction with money is, however, mainly a "hygiene factor," as Frederick Herzberg called it all of forty years ago in his 1959 book *The Motivation to Work*. What motivates—and especially what motivates knowledge workers—is what motivates volunteers. Volunteers, we know, have to get *more* satisfaction from their work than paid employees, precisely because they do not get a paycheck. They need, above all, challenge. They need to know the organization's mission and to believe in it. They need continual training. They need to see results.

Implicit in this is that different groups in the work population have to be managed differently, and that the same group in the work population has to be managed differently at different times. Increasingly "employees" have to be managed as "partners"—and it is the definition of a partnership that all partners are equal. It is also the definition of a partnership that partners cannot be ordered. They have to be persuaded. Increasingly, therefore, the management of people is a "marketing job." And in marketing one does not begin with the question, What do *we* want? One begins with the questions, What does the other party want? What are its values?

What are its goals? What does it consider results? And this is neither "Theory X" nor "Theory Y," nor any other specific theory of *managing* people.

Maybe we will have to redefine the task altogether. It may not be "managing the work of people." The starting point both in theory and in practice may have to be "managing for performance." The starting point may be a definition of results—just as the starting points of both the orchestra conductor and the football coach are the score.

The productivity of the knowledge worker is likely to become the center of the management of people, just as the work on the productivity of the manual worker became the center of managing people a hundred years ago, that is, since Frederick W. Taylor. This will require, above all, very different assumptions about people in organizations and their work:

One does not "manage" people.
The task is to lead people.
And the goal is to make productive the specific strengths and knowledge of each individual.

Technologies and End Uses Are Fixed and Given

Four major assumptions, as stated above, have been underlying the <u>practice</u> of management all along—in fact for much longer than there has been a <u>discipline</u> of management.

The assumptions about technology and end uses to a very large extent underlie the rise of modern business and of the modern economy altogether. They go back to the very early days of the Industrial Revolution.

When the textile industry first developed out of what had been cottage industries, it was assumed—and with complete validity—that the textile industry had its own unique technology. The same was true in respect to coal mining, and of any of the other industries that arose in the late eighteenth century and the first half of the nineteenth century. The first one to understand this and to base a major enterprise on it was also one of the first men to develop what we would today call a modern business, the German Werner Siemens (1816–1892). It led him in 1869 to hire the first university-trained scientist to start a modern research lab—devoted exclusively to what we would now call electronics, and based on a clear under-standing that electronics (in those days called "low-voltage") was distinct and separate from all other industries, and had its distinct and separate technology.

Out of this insight grew not only Siemens's own company with its own research lab, but also the German chemical industry, which assumed worldwide leadership because it based itself on the assumption that chemistry—and especially organic chemistry—had its own unique technology. Out of it then grew all the other major companies the world over, whether the American electrical and chemical companies, the automobile companies, the telephone companies, and so on. Out of this insight then grew what may well be the most successful invention of the nineteenth century, the research laboratory—the last one almost a century after Siemens's, the 1950 lab of IBM—and at around the same time the research labs of the major pharmaceutical companies as they emerged as a worldwide industry after World War II.

By now this assumption has become untenable. The best exam-ple is of course the pharmaceutical industry, which increasingly has come to depend on technologies that are fundamentally different from the technologies on which the pharmaceutical research lab is based: genetics, for instance, microbiology, molecular biology, medical electronics, and so on.

In the nineteenth century and throughout the first half of the twentieth century, it could be taken for granted that technologies outside one's own industry had no, or at least only minimal, impact on the industry. Now the assumption to start with is that the tech-nologies that are likely to have the greatest impact on a company and an industry are technologies outside its own field.

The original assumption was of course that one's own research lab would and could produce everything the company—or the company's industry—needed. And in turn the assumption was that everything that this research lab produced would be used in and by the industry that it served.

This, for instance, was the clear foundation of what was probably the most successful of all the great research labs of the last hundred years, the Bell Labs of the American telephone system. Founded in the early 1920s, the Bell Labs until the late 1960s did indeed produce practically every new knowledge and every new technology the telephone industry needed. And in turn practically everything the Bell Labs scientists produced found its main use in the telephone system. This changed drastically with what was probably the Bell Labs' greatest scientific achievement: the transistor. The telephone company itself did become a heavy user of the transistor. But the main uses of the transistor were outside the telephone system. This was so unexpected that the Bell Telephone Company, when the transistor was first developed, virtually gave it away—it did not see enough use for it within the telephone system. But it also did not see any use for it outside it. And so what was the most revolutionary development that came out of the Bell Labs—and certainly the most valuable one—was sold freely to all comers for the paltry sum of twenty-five thousand dollars. It is on this total failure of the Bell Labs to understand the significance of its own achievement that practically all modern electronics companies outside of the telephone are based.

Conversely, the things that have revolutionized the telephone system—such as digital switching or the fiberglass cable—did not come out of the Bell Labs. They came out of technologies that were foreign to telephone technology. And this has been typical altogether of the last thirty to fifty years—and it is increasingly becoming more typical of every industry.

Today's technologies, unlike those of the nineteenth century, no longer run in parallel lines. They constantly crisscross. Constantly, something in a technology of which people in a given industry have barely heard (just as the people in the pharmaceutical industry had never heard of genetics, let alone of medical electronics) revolutionizes an industry and its technology. Constantly, such outside technologies force an industry to learn, to acquire, to adapt, to change its very mind-set, let alone its technical knowledge.

Equally important to the rise of nineteenth- and early-twentieth-century industry and companies was a second assumption: End uses are fixed and given. For a certain end use, for example, to put beer into containers, there may have been extreme competition between various suppliers of containers. But all of them, until recently, were glass companies, and there was only one way of putting beer into containers, a glass bottle.

This was accepted as obvious not only by business, industry, and the consumer, but by governments as well. The American regulation of business rests on the assumptions that to every industry pertains a unique technology and that to every end use pertains a specific and unique product or service. These are the assumptions on which antitrust legislation was based. And to this day antitrust advocates concern themselves with the domination of the market in glass bottles and pay little attention to the fact that beer increasingly is not put into glass bottles but into cans (or, vice versa, they concern themselves exclusively with the concentration of supply in respect to metal containers for beer, paying no attention to the fact that beer is still being put into glass bottles, but also increasingly into plastic cans).

But since World War II, end uses are not uniquely tied anymore to a certain product or service. The plastics of course were the first major exception to the rule. But by now it is clear that it is not just one material moving in on what was considered the "turf" of another one. Increasingly, the same want is being satisfied by very different means. It is the *want* that is unique, and not the means to satisfy it.

As late as the beginning of World War II, news was basically the monopoly of the newspaper—an eighteenth-century invention that saw its biggest growth in the early years of the twentieth century. By now there are several competing ways to deliver news: still the printed newspaper; increasingly, the same newspaper delivered online through the Internet; radio; television; separate news organizations that use only electronics—as is increasingly the case with economic and business news—and quite a few additional ones.

And then there is the new "basic resource" information. It differs radically from all other commodities in that it does not stand under the scarcity theorem. On the contrary, it stands under an abundance theorem. If I sell a thing—for example, a book—I no longer have the book. If I impart information, I still have it. And in fact, information becomes more valuable the more people have it. What this means for economics is well beyond the scope of this book—though it is clear that it will force us radically to revise basic economic theory. But it also means a good deal for management. Increasingly basic assumptions will have to be changed. Information does not pertain to any industry or to any business. Information also does not have any one end use, nor does any end use require a particular kind of information or depend on one particular kind of information.

Management therefore now has to start out with the assumption that there is no one technology that pertains to any industry and that, on the contrary, all technologies are capable—and indeed likely—to be of major importance to any industry and to have impact on any industry. Management similarly has to start with the assumption that there is no one given end use for any product or service and that, conversely, no end use is going to be linked to any one product or service.

Some implications of this are that increasingly the *noncustomers* of an enterprise—whether a business, a university, a church, a hospital—are as important as the customers, if not more important.

Even the biggest enterprise (other than a government monopoly) has many more noncustomers than it has customers. There are very few institutions that supply as large a percentage of a market as 30 percent. There are therefore few institutions where the noncustomers do not amount to at least 70 percent of the potential market. And yet very few institutions know anything about the noncustomers—very few of them even know that they exist, let alone know who they are. And even fewer know why they are not customers. Yet it is with the noncustomers that changes always start.

Another critical implication is that the starting point for management can no longer be its own product or service, and not even its known market and its known end uses for its products and services. The starting point has to be what *customers consider value.* The starting point has to be the assumption—an assumption amply proven by all our experience—that the customer never buys what the supplier sells. What is value to the customer is always something quite different from what is value or quality to the supplier. This applies as much to a business as to a university or to a hospital.

Management, in other words, will increasingly have to be based on the assumption that neither technology nor end use is a foundation for management policy. They are limitations. *The foundations have to be customer values and customer decisions on the distribution of their disposable income. It is with those that management policy and management strategy increasingly will have to start.*

Management's Scope Is Legally Defined

Management, both in theory and in practice, deals with the legal entity, the individual enterprise—whether the business corporation, the hospital, the university, and so on. The scope of management is thus *legally* defined. This has been—and still is—the almost universal assumption.

One reason for this assumption is the traditional concept of management as being based on command and control. Command and control are indeed legally defined. The chief executive of a business, the bishop of a diocese, the administrator of a hospital, have no command and control authority beyond the legal confines of their institution.

Almost a hundred years ago, it first became clear that the legal definition was not adequate to manage a major enterprise.

The Japanese are usually credited with the invention of the "keiretsu," the management concept in which the suppliers to an enterprise are tied together with their main customer, for example, Toyota, for planning, product development, cost control, and so on. But actually the keiretsu is much older and an American invention. It goes back to around 1910 and to the man who first saw the potential of the automobile to become a major industry, William C. Durant (1861–1947). It was Durant who created General Motors by buying up small but successful automobile manufacturers such as Buick and merging them into one big automobile company. A few years later Durant then realized that he needed to bring the main suppliers into his corporation. He began to buy up and merge into General Motors one parts and accessories maker after the other, finishing in 1920 by buying Fisher Body, the country's largest manufacturer of automobile bodies. With this purchase General Motors had come to own the manufacturers of 70 percent of everything that went into its automobiles—and had become by far the world's most integrated large business. It was this prototype keiretsu that gave General Motors the decisive advantage, both in cost and in speed, which made it within a few short years both the world's largest and the world's most profitable manufacturing company, and the unchallenged leader in an exceedingly competitive American automobile market. In fact, for some thirty-odd years, General Motors enjoyed a 30 percent cost advantage over all its competitors, including Ford and Chrysler.

But the Durant keiretsu was still based on the belief that management means command and control—this was the reason that Durant *bought* all the companies that became part of General Motors's keiretsu. And this eventually became the greatest weakness of GM. Durant had carefully planned to ensure the competitiveness of the GM-owned accessory suppliers. Each of them (except Fisher Body) had to sell 50 percent of its output outside of GM, that is, to competing automobile manufacturers, and thus had to maintain competitive costs and competitive quality. But after World War II the competing automobile manufacturers disappeared—and with them the check on the competitiveness of GM's wholly owned accessory divisions. Also, with the unionization of the automobile industry in 1936–1937, the high labor costs of automobile assembly plants were imposed on General Motors's accessory divisions, which put them at a cost disadvantage that to this day they have not been able to overcome. That Durant based his keiretsu on the assumption that management means command and control largely explains, in other words, the decline of General Motors in the last twenty-five years and the company's inability to turn itself around.

This was clearly realized in the 1920s and 1930s by the builder of the next keiretsu, Sears, Roebuck. As Sears became America's largest retailer, especially of appliances and hardware, it too realized the necessity of bringing together into one group its main suppliers so as to make possible joint planning, joint product development and product design, and cost control across the entire economic chain. But instead of buying these suppliers, Sears bought small minority stakes in them—more as a token of its commitment than as an investment—and based the relationship otherwise on contract. And the next keiretsu builder—and probably the most successful one so far (even more successful than the Japanese)—was Marks & Spencer in England, which, beginning in the early 1930s, integrated practically all its suppliers into its own management system, but exclusively through contracts rather than through ownership stakes or ownership control.

It is the Marks & Spencer model that the Japanese, quite consciously, copied in the 1960s.

In every case, beginning with General Motors, the keiretsu—that is, the integration into one management system of enterprises that are linked economically rather than controlled legally—has given a cost advantage of at least 25 percent and more often 30 percent. In every case, it has given dominance in the industry and in the marketplace.

And yet the keiretsu is not enough. It is still based on power. Whether it is General Motors and the small, independent accessory companies that Durant bought between 1915 and 1920; or Sears, Roebuck; or Marks & Spencer; or Toyota—the central company has overwhelming economic power. The keiretsu is not based on a partnership of equals. It is based on the dependence of the suppliers.

Increasingly, however, the economic chain brings together genuine *partners,* that is, institutions in which there is equality of power and genuine independence. This is true of the partnership between a pharmaceutical company and the biology faculty of a major research university. It is true of the joint ventures through which American industry got into Japan after World War II. It is true of the partnerships today between chemical and pharmaceutical companies and companies in genetics, molecular biology, or medical electronics.

These companies in the new technologies may be quite small— and very often are—and badly in need of capital. But they own independent technology. Therefore *they* are the senior partners when it comes to technology. They, rather than the much bigger pharmaceutical or chemical company, have a choice of with whom to ally themselves. The same is largely true in information technology, and also in finance. And then neither the traditional keiretsu nor command and control work.

What is needed, therefore, is a redefinition of the scope of management. *Management has to encompass the entire process.* For business this means by and large the economic process.

The new assumption on which management, both as a discipline and as a practice, will increasingly have to base itself is that the scope of management is not legal.

It has to be *operational. It has to embrace the entire process. It has to be focused on results and performance across the entire economic chain.*

Management's Scope Is Politically Defined

It is still generally assumed in the discipline of management—and very largely still taken for granted in the practice of management—that the domestic economy, as defined by national boundaries, is the ecology of enterprise and management—and of nonbusinesses as much as of businesses.

This assumption underlies the traditional "multinational."

As is well known, before World War I, as large a share of the world's production of manufactured goods and of financial services was multinational as it is now. The 1913 leading company in any industry, whether in manufacturing or in finance, derived as large a share of its sales from selling outside its own country as it did by selling inside its own country. But insofar as it produced outside its own national boundaries, it produced within the national boundaries of another country.

One example:

The largest supplier of war matériel to the Italian Army during World War I was a young but rapidly growing company called Fiat in Turin—it made all the automobiles and trucks for the Italian army. The largest supplier of war matériel to the Austro-Hungarian army in World War I was also a company called Fiat—in Vienna. It supplied all the automobiles and trucks to the Austro-Hungarian army. It was two to three times the size of its parent company. For Austria-Hungary was a much larger market than Italy, partly because it had a much larger population, and partly because it was more highly developed, especially in its western parts. Fiat-Austria was wholly owned by Fiat-Italy. But except for the designs that came from Italy, Fiat-Austria was a separate company. Everything it used was made or bought in Austria. All products were sold in

Austria. And every employee up to and including the CEO was an Austrian. When World War I came, and Austria and Italy became enemies, all the Austrians had to do, therefore, was to change the bank account of Fiat-Austria—it kept on working as it had all along.

Even traditional industries like the automotive industry or insurance are no longer organized that way.

Post–World War II industries such as the pharmaceutical industry, or the information industries, are increasingly not even organized in "domestic" and "international" units as GM and Allianz still are. They are run as a worldwide system in which individual tasks, whether research, design, engineering, development, testing and increasingly manufacturing and marketing, are each organized "transnationally."

One large pharmaceutical company has seven labs in seven different countries, each focusing on one major area (e.g., antibiotics) but all run as one "research department" and all reporting to the same research director in headquarters. The same company has manufacturing plants in eleven countries, each highly specialized and producing one or two major product groups for worldwide distribution and sale. It has one medical director who decides in which of five or six countries a new drug is to be tested. But managing the company's foreign exchange exposure is totally centralized in one location for the entire system.

In the traditional multinational, economic reality and political reality were congruent. The country was the "business unit," to use today's term. In today's transnational—but increasingly, also, in the old multinationals as they are being forced to transform themselves—the country is only a "cost center." It is a complication rather than the unit for organization and the unit of business, of strategy, of production, and so on.

Management and national boundaries are no longer congruent. The scope of management can no longer be politically defined. National boundaries will continue to be important.

But the new assumption has to be:

National boundaries are important primarily as restraints. The practice of management—and by no means for businesses only—will increasingly have to be defined *operationally* rather than *politically.*

The Inside Is Management's Domain

All the traditional assumptions led to one conclusion: *The inside of the organization is the domain of management.*

This assumption explains the otherwise totally incomprehensible distinction between management and entrepreneurship.

In actual practice this distinction makes no sense whatever. An enterprise, whether a business or any other institution, that does not innovate and does not engage in entrepreneurship will not survive long.

It should have been obvious from the beginning that management and entrepreneurship are only two different dimensions of the same task. An entrepreneur who does not learn how to manage will not last long. A management that does not learn to innovate will not last long. In fact, business—and every other organization today—has to be designed for change as the norm and to create change rather than react to it.

But entrepreneurial activities start with the outside and are focused on the outside. They therefore do not fit within the traditional assumptions of management's domain—which explains why they have come so commonly to be regarded as different, if not incompatible. Any organization, however, that actually believes that management and entrepreneurship are different, let alone incompatible, will soon find itself out of business.

The inward focus of management has been greatly aggravated in the last decades by the rise of information technology. Information technology so far may actually have done more damage to management than it has helped.

The traditional assumption that the inside of the organization is the domain of management means that management is assumed to concern itself with *efforts,* if not with *costs* only. For effort is the only thing that exists within an organization. And, similarly, everything inside an organization is a cost center.

But results of any institution exist only on the outside.

It is understandable that management *began* as a concern for the inside of the organization. When the large organizations first arose—with the business enterprise, around 1870, the first and by far the most visible one—managing the inside was the new challenge. Nobody had ever done it before. But while the assumption that management's domain is the inside of the organization originally made sense—or at least can be explained—its continuation makes no sense whatever. It is a contradiction of the very function and nature of organization.

Management must focus on the *results* and *performance* of the organization. Indeed, the first task of management is to define what results and performance are in a given organization—and this, as anyone who has worked on it can testify, is in itself one of the most difficult, one of the most controversial, but also one of the most important tasks. It is therefore the specific function of management to organize the resources of the organization *for results outside the organization.*

The new assumption—and the basis for the new paradigm on which management, both as a discipline and as a practice, has to be based—is therefore:

Management exists for the sake of the institution's results. It has to start with the intended results and has to organize the resources of the institution to attain these results. It is the organ to make the institution, whether business, church, university, hospital, or a battered women's shelter, capable of producing results outside of itself.

This chapter has not tried to give answers—intentionally so. It has tried to raise questions. But underlying all of these is one insight. The center of a modern society, economy, and community is not technology. It is not information. It is not productivity. *It is the managed institution as the organ of society to produce results.* And management is the specific tool, the specific function, the specific instrument to make institutions capable of producing results.

This, however, requires a <u>final</u> new management paradigm.

Management's concern and management's responsibility are everything that affects the performance of the institution and its results—whether inside or outside, whether under the institution's control or totally beyond it.

7.

THE INFORMATION
EXECUTIVES NEED TODAY

Ever since the new data-processing tools first emerged thirty or forty years ago, businesspeople have both over- and underrated the importance of information in the organization. We—and I include myself—overrated the possibilities to the point where we talked of computer-generated "business models" that could make decisions and might even be able to run much of the business. But we also grossly underrated the new tools; we saw in them the means to do better what executives were already doing to manage their organizations.

Nobody talks anymore of business models making economic decisions. The greatest contribution of our data-processing capacity so far has not even been to management. It has been to operations—in the form of such things as the marvelous software that architects now use to solve structural problems in the buildings they design.

Yet even as we both over- and underestimated the new tools, we failed to realize that they would drastically change the *tasks* to be tackled. Concepts and tools, history teaches again and again, are mutually interdependent and interactive. One changes the other.

That is now happening to the concept we call a business and to the tools we use to collect information. The new tools enable us—indeed, may force us—to see our businesses differently, to see them as

> ➤ Generators of resources, that is, as the organizations that convert costs into yields
> ➤ Links in an economic chain, which managers need to understand as a whole in order to manage their costs
> ➤ Society's organs for the creation of wealth
> ➤ Both creators and creatures of a material environment, which is the area outside the organization in which opportunities and results lie but in which the threats to the success and survival of every business also originate

This chapter deals with the tools executives require to generate the information they need. And it deals with the concepts underlying those tools. Some of the tools have been around for a long time, but rarely, if ever, have they been focused on the task of managing a business. Some have to be refashioned; in their present form they no longer work. For some tools that promise to be important in the future, we have so far only the briefest specifications. The tools themselves still have to be designed.

Even though we are just beginning to understand how to use information as a tool, we can outline with high probability the major parts of the information system executives need to manage their businesses. So, in turn, can we begin to understand the concepts likely to underlie the business—call it the redesigned corporation—that executives will have to manage tomorrow.

From Cost Accounting to Yield Control

We may have gone furthest in redesigning both business and information in the most traditional of our information systems: accounting. In fact, many businesses have already shifted from traditional cost accounting to activity-based costing. Activity-based costing represents both a different concept of the business process, especially for manufacturers, and different ways of measuring.

Traditional cost accounting, first developed by General Motors seventy years ago, postulates that total manufacturing cost is the sum of the costs of individual operations. Yet the cost that matters for competitiveness and profitability is the cost of the total process, and that is what the new *activity-based costing* records and makes manageable. Its basic premise is that manufacturing is an integrated process that starts when supplies, materials, and parts arrive at the plant's loading dock and continues even after the finished product reaches the end user. Service is still a cost of the product, and so is installation, even if the customer pays.

Traditional cost accounting measures what it costs to *do* something, for example to cut a screw thread. Activity-based costing also records the cost of *not doing,* such as the cost of machine downtime, the cost of waiting for a needed part or tool, the cost of inventory waiting to be shipped, and the cost of reworking or scrapping a defective part. The costs of not doing, which traditional cost accounting cannot and does not record, often equal and sometimes even exceed the costs of doing. Activity-based costing therefore not only gives much better cost control but, increasingly, it also gives *result control.*

Traditional cost accounting assumes that a certain operation— for example, heat treating—has to be done and that it has to be done where it is being done now. Activity-based costing asks, Does it have to be done? If so, where is it best done? Activity-based costing integrates what were once several activities—value analysis, process analysis, quality management, and costing—into one analysis.

Using that approach, activity-based costing can substantially lower manufacturing costs—in some instances by a full third or more. Its greatest impact, however, is likely to be in services. In most manufacturing companies, cost accounting is inadequate. But service industries—banks, retail stores, hospitals, schools, newspapers, and radio and television stations—have practically no cost information at all.

Activity-based costing shows us why traditional cost accounting has not worked for service companies. It is not because the techniques are wrong. It is because traditional cost accounting makes the wrong assumptions. Service companies cannot start with the cost of individual operations, as manufacturing companies have done with traditional cost accounting. They must start with the assumption that there is only *one* cost: that of the total system. And it is a fixed cost over any given time period. The famous distinction between fixed and variable costs, on which traditional cost accounting is based, does not make much sense in services. Neither does another basic assumption of traditional cost accounting: that capital can be substituted for labor. In fact, in knowledge-based work especially, additional capital investment will likely require more rather than less labor. For example, a hospital that buys a new diagnostic tool may have to add four or five people to run it. Other knowledge-based organizations have had to learn the same lesson.

But that all costs are fixed over a given time period and that resources cannot be substituted for one another, so that the *total* operation has to be costed—those are precisely the assumptions with which activity-based costing starts. By applying them to services, we are beginning for the first time to get cost information and yield control.

Banks, for instance, have been trying for several decades to apply conventional cost-accounting techniques to their business— that is, to figure the costs of individual operations and services— with almost negligible results. Now they are beginning to ask Which one *activity* is at the center of costs and of results? The answer: serving the customer. The cost per customer in any major area of banking is a fixed cost. Thus it is the *yield* per customer— both the volume of services a customer uses and the mix of those services—that determines costs and profitability. Retail discounters, especially those in Western Europe, have known that for some time. They assume that once a unit of shelf space is installed, the cost is fixed and management consists of maximizing the yield thereon over a given time span. Their focus on yield control has enabled them to increase profitability despite their low prices and low margins.

Services are still only beginning to apply the new costing concepts. In some areas, such as research labs, where productivity is nearly impossible to measure, we may always have to rely on assessment and judgment rather than on measurement. But for most knowledge-based and service work, we should, within ten to fifteen years, have developed reliable tools to measure and manage costs and to relate those costs to results.

Thinking more clearly about costing in services should yield new insights into the costs of getting and keeping customers in businesses of all kinds. If GM, Ford, and Chrysler had had activity-based costing, for example, they would have realized early on the utter futility of their competitive blitzes of the past few years, which offered new-car buyers spectacular discounts and hefty cash rewards. Those promotions actually cost the Big Three carmakers enormous amounts of money and, worse, enormous numbers of potential customers.

From Legal Fiction to Economic Reality

Knowing the cost of your operations, however, is not enough. To compete successfully in an increasingly competitive global market, a company has to know the costs of its entire economic chain and has to work with other members of the chain to manage costs and maximize yield. Companies are therefore beginning to shift from costing only what goes on inside their own organizations to costing the entire economic process, in which even the biggest company is just one link.

The legal entity, the company, is a reality for shareholders, for creditors, for employees, and for tax collectors. But *economically,* it is fiction. Thirty years ago the Coca-Cola Company was a franchisor. Independent bottlers manufactured the product. Now the company owns most of its bottling operations in the United States. But Coke drinkers—even those few who know that fact—could not care less. What matters in the marketplace is the economic reality, the costs of the entire process, regardless of who owns what.

Again and again in business history, an unknown company has come from nowhere and in a few short years overtaken the established leaders without apparently even breathing hard. The explanation always given is superior strategy, superior technology, superior marketing, or lean manufacturing. But in every single case, the newcomer also enjoys a tremendous cost advantage, usually about 30 percent. The reason is always the same: the new company knows and manages the costs of the entire economic chain rather than its costs alone.

Toyota is perhaps the best-publicized example of a company that knows and manages the costs of its suppliers and distributors; they are all, of course, members of its keiretsu. Through that network, Toyota manages the total cost of making, distributing, and servicing its cars as one cost stream, putting work where it costs the least and yields the most.

Managing the economic cost stream is not a Japanese invention, however, but an American one. It began with the man who designed and built General Motors, William Durant. In about 1908, Durant began to buy small, successful car companies—Buick, Oldsmobile, Cadillac, Chevrolet—and merged them into his new General Motors Corporation. In 1916, he set up a separate subsidiary called United Motors to buy small, successful parts companies. His first acquisitions included Delco, which held Charles Kettering's patents to the automotive self-starter.

Durant ultimately bought about twenty supplier companies; his last acquisition—in 1919, the year before he was ousted as GM's CEO—was Fisher Body. Durant deliberately brought the parts and accessories makers into the process of designing a new car model right from the start. Doing so allowed him to manage the total costs of the finished car as one cost stream. And in so doing, Durant invented the keiretsu.

However, between 1950 and 1960, Durant's keiretsu became an albatross around the company's neck. Unionization imposed higher labor costs on GM's parts divisions than on their independent competitors. The outside customers, the independent car companies such as Packard and Studebaker, which had bought 50 percent of the output of GM's parts divisions, disappeared one by one. And GM's control over both the costs and quality of its main suppliers disappeared with them. Nevertheless, for more than forty years, GM's systems costing gave it an unbeatable advantage over even the most efficient of its competitors, which for most of that time was Studebaker.

Sears, Roebuck and Company was the first to copy Durant's system. In the 1920s it established long-term contracts with its suppliers and bought minority interests in them. Sears was then able to consult with suppliers as it designed the product and to understand and manage the entire cost stream. That gave the company an unbeatable cost advantage for decades.

In the early 1930s, London-based Marks & Spencer copied Sears, with the same result. Twenty years later, the Japanese, led by Toyota, studied and copied both Sears and Marks & Spencer. Then in the 1980s, Wal-Mart Stores adapted the approach by allowing suppliers to stock products directly on store shelves, thereby eliminating warehouse inventories and with them nearly one-third of the cost of traditional retailing.

But those companies are still exceptions. Although economists have known the importance of costing the entire economic chain since Alfred Marshall wrote about it in the late 1890s, most businesspeople still consider it theoretical abstraction. Increasingly, however, managing the economic cost chain will become a necessity. Indeed, executives need to organize and manage not only the cost chain but also everything else—especially corporate strategy and product planning—as one economic whole, regardless of the legal boundaries of individual companies.

A powerful force driving companies toward economic-chain costing will be the shift from cost-led pricing to price-led costing. Traditionally, Western companies have started with costs, put a desired profit margin on top, and arrived at a price. They practiced cost-led pricing. Sears and Marks & Spencer long ago switched to price-led costing, in which the price the customer is willing to pay determines allowable costs, beginning with the design stage. Until recently, those companies were the exceptions. Now price-led costing is becoming the rule. The Japanese first adopted it for their exports. Now Wal-Mart and all the discounters in the United States, Japan, and Europe are practicing price-led costing. It underlies Chrysler's success with its recent models and the success of GM's Saturn. Companies can practice price-led costing, however, only if they know and manage the *entire* cost of the economic chain.

The same ideas apply to outsourcing, alliances, and joint ventures—indeed, to any business structure that is built on partnership rather than control. And such entities, rather than the traditional model of a parent company with wholly owned subsidiaries, are increasingly becoming the models for growth, especially in the global economy.

Still, it will be painful for most businesses to switch to economic-chain costing. Doing so requires uniform or at least compatible accounting systems at companies along the entire chain. Yet each one does its accounting in its own way, and each is convinced that its system is the only possible one. Moreover, economic-chain costing requires information sharing across companies, yet even within the same company, people tend to resist information sharing. Despite those challenges, companies can find ways to practice economic-chain costing now, as Procter & Gamble is demonstrating. Using the way Wal-Mart develops close relationships with suppliers as a model, P&G is initiating information sharing and economic-chain management with the three hundred large retailers that distribute the bulk of its products worldwide.

Whatever the obstacles, economic-chain costing is going to be done. Otherwise, even the most efficient company will suffer from an increasing cost disadvantage.

Information for Wealth Creation

Enterprises are paid to create wealth, not to control costs. But that obvious fact is not reflected in traditional measurements. First-year accounting students are taught that the balance sheet portrays the liquidation value of the enterprise and provides creditors with worst-case information. But enterprises are not normally run to be liquidated. They have to be managed as going concerns, that is, for *wealth creation*. To do that requires information that enables executives to make informed judgments. It requires four sets of diagnostic tools: foundation information, productivity information, competence information, and information about the allocation of scarce resources. Together, they constitute the executive's tool kit for managing the current business.

The oldest and most widely used set of diagnostic management tools are cash-flow and liquidity projections and such standard measurements as the ratio between dealers' inventories and sales of new cars; the earnings coverage for the interest payments on a bond issue; and the ratios among receivables outstanding more than six months, total receivables, and sales. Those may be likened to the measurements a doctor takes at a routine physical: weight, pulse, temperature, blood pressure, and urine analysis. If those readings are normal, they do not tell us much. If they are abnormal, they indicate a problem that needs to be identified and treated. Those measurements might be called foundation information.

The second set of tools for business diagnosis deals with the productivity of key resources. The oldest of them—of World War II vintage—measures the productivity of manual labor. Now we are slowly developing measurements, though still quite primitive ones, for the productivity of knowledge-based and service work. However, measuring only the productivity of workers, whether blue- or white-collar, no longer gives us adequate information about productivity. We need data on *total-factor productivity*.

That explains the popularity of economic value-added analysis. EVA is based on something we have known for a long time: what we generally call profits, the money left to service equity, is usually not profit at all. Until a business returns a profit that is greater than its cost of capital, it operates at a loss. Never mind that it pays taxes as if it had a genuine profit. The enterprise still returns less to the economy than it devours in resources. It does not cover its full costs unless the reported profit exceeds the cost of capital. Until then, it does not create wealth; it destroys it. By that measurement, incidentally, few U.S. businesses have been profitable since World War II.

By measuring the value added over *all* costs, including the cost of capital, EVA measures, in effect, the productivity of *all* factors of production. It does not, by itself, tell us why a certain product or a certain service does not add value or what to do about it. But it shows us what we need to find out and whether we need to take remedial action. EVA should also be used to determine what works. It does show which product, service, operation, or activity has unusually high productivity and adds unusually high value. Then we should ask ourselves, What can we learn from those successes?

The most recent of the tools used to obtain productivity information is benchmarking—comparing one's performance with the best performance in the industry or, better yet, with the best anywhere in business. Benchmarking assumes correctly that what one organization does, any other organization can do as well. And it assumes, also correctly, that being at least as good as the leader is a prerequisite to being competitive. Together, EVA and benchmarking provide the diagnostic tools to measure total-factor productivity and to manage it.

A third set of tools deals with competences. Ever since C. K.

Prahalad and Gary Hamel's pathbreaking article "The Core Competence of the Corporation" (*Harvard Business Review,* May–June 1990), we have known that leadership rests on being able to do something others cannot do at all or find difficult to do even poorly. It rests on core competencies that meld market or customer value with a special ability of the producer or supplier.

Some examples: the ability of the Japanese to miniaturize electronic components, which is based on their three-hundred-year-old artistic tradition of putting landscape paintings on a tiny lacquered box, called an *inro,* and of carving a whole zoo of animals on the even tinier button, called a *netsuke,* that holds the box on the wearer's belt; or the almost unique ability GM has had for eighty years to make successful acquisitions; or Marks & Spencer's also unique ability to design packaged and ready-to-eat gourmet meals for middle-class purses. But how does one identify both the core competencies one has already and those the business needs in order to take and maintain a leadership position? How does one find out whether one's core competence is improving or weakening? Or whether it is still the right core competence and what changes it might need?

So far, the discussion of core competencies has been largely anecdotal. But a number of highly specialized midsize companies— a Swedish pharmaceuticals producer and a U.S. producer of specialty tools, to name two—are developing the methodology to measure and manage core competencies. The first step is to keep careful track of one's own and one's competitors' performances, looking especially for unexpected successes and for unexpected poor performance in areas where one should have done well. The successes demonstrate what the market values and will pay for. They indicate where the business enjoys a leadership advantage. The nonsuccesses should be viewed as the first indication either that the market is changing or that the company's competencies are weakening.

That analysis allows for the early recognition of opportunities. For example, by carefully tracking an unexpected success, a U.S. toolmaker found that small Japanese machine shops were buying its high-tech, high-priced tools, even though it had not designed the tools with them in mind or even called on them. That allowed the company to recognize a new core competence: the Japanese were attracted to its products because they were easy to maintain and repair despite their technical complexity. When that insight was applied to designing products, the company gained leadership in the small-plant and machine-shop markets in the United States and Western Europe, huge markets where it had done practically no business before.

Core competencies are different for every organization; they are, so to speak, part of an organization's personality. But every organization—not just businesses—needs one core competence: *innovation*. And every organization needs a way to record and appraise its *innovative performance*. In organizations already doing that—among them several top-flight pharmaceuticals manufacturers—the starting point is not the company's own performance. It is a careful record of the innovations in the entire field during a given period. Which of them were truly successful? How many of them were ours? Is our performance commensurate with our objectives? With the direction of the market? With our market standing? With our research spending? Are our successful innovations in the areas of greatest growth and opportunity? How many of the truly important innovation opportunities did we miss? Why? Because we did not see them? Or because we saw them but dismissed them? Or because we botched them? And how well do we convert an innovation into a commercial product? A good deal of that, admittedly, is assessment rather than measurement. It raises rather than answers questions, but it raises the right questions.

The last area in which diagnostic information is needed to manage the current business for wealth creation is the allocation of scarce resources: capital and performing people. Those two convert into action whatever information management has about its business. They determine whether the enterprise will do well or poorly.

GM developed the first systematic capital-appropriations process about seventy years ago. Today practically every business has a capital-appropriations process, but few use it correctly. Companies typically measure their proposed capital appropriations by only one or two of the following four yardsticks: return on investment, pay-back period, cash flow, and discounted present value. But we have known for a long time—since the early 1930s—that none of those is *the* right method. To understand a proposed investment, a company needs to look at *all four*. Sixty years ago that would have required endless number-crunching. Now a laptop computer can provide the information within a few minutes. We also have known for sixty years that managers should never look at just one proposed capital appropriation in isolation but should instead choose the projects that show the best ratio between opportunity and risks. That requires a capital-appropriations *budget* to display the choices—again, something far too many businesses do not do. Most serious, however, is that most capital-appropriations processes do not even ask for two vital pieces of information:

- ➤ What will happen if the proposed investment fails to produce as promised as do three out of every five? Would it seriously hurt the company, or would it be just a flea bite?
- ➤ If the investment is successful—and especially if it is more successful than we expect—what will it commit us to?

No one at GM seemed to have asked what Saturn's success would commit the company to. As a result, the company may end up killing its own success because of its inability to finance it.

In addition, a capital-appropriations request requires specific deadlines: when should we expect what results? Then the results—successes, near-successes, near-failures, and failures—need to be reported and analyzed. There is no better way to improve an organization's performance than to measure the results of capital appropriations against the promises and expectations that led to their authorization. How much better off the United States would be today had such feedback on government programs been standard practice for the past fifty years.

Capital, however, is only one key resource of the organization, and it is by no means the scarcest one. The scarcest resources in any organization are performing people. Since World War II, the U.S. military—and so far no one else—has learned to test its placement decisions. It now thinks through what it expects of senior officers before it puts them into key commands. It then appraises their performance against those expectations. And it constantly appraises its own process for selecting senior commanders against the successes and failures of its appointments. In business, by contrast, placement with specific expectations as to what the appointee should achieve and systematic appraisal of the outcome are virtually unknown. In the effort to create wealth, managers need to allocate human resources as purposefully and as thoughtfully as they do capital. And the outcomes of those decisions ought to be recorded and studied as carefully.

Where the Results Are

The four kinds of information tell us only about the current business. They inform and direct *tactics*. For *strategy*, we need organized information about the environment. Strategy has to be based on information about markets, customers, and noncustomers; about technology in one's own industry and others; about worldwide finance; and about the changing world economy. For that is where the results are. Inside an organization, there are only cost centers. The only profit center is a customer whose check has not bounced.

Major changes also start outside an organization. A retailer may know a great deal about the people who shop at its stores. But no matter how successful it is, no retailer ever has more than a small fraction of the market as its customers; the great majority are noncustomers. It is always with noncustomers that basic changes begin and become significant.

At least half the important new technologies that have transformed an industry in the past fifty years came from outside the industry itself. Commercial paper, which has revolutionized finance in the United States, did not originate with the banks. Molecular biology and genetic engineering were not developed by the pharmaceuticals industry. Though the great majority of businesses will continue to operate only locally or regionally, they all face, at least potentially, global competition from places they have never even heard of before.

Not all the needed information about the outside is available, to be sure. There is no information—not even unreliable information—on economic conditions in most of China, for instance, or on legal conditions in most of the successor states to the Soviet Empire. But even where information is readily available, many businesses are oblivious to it. Many U.S. companies went into Europe in the 1960s without even asking about labor legislation. European companies have been just as blind and ill informed in their ventures into the United States. A major cause of the Japanese real estate investment debacle in California during the 1990s was the failure to find out elementary facts about zoning and taxes.

A serious cause of business failure is the common assumption that conditions—taxes, social legislation, market preferences, distribution channels, intellectual property rights, and many others—*must* be what we think they are or at least what we think they *should be*. An adequate information system has to include information that makes executives question that assumption. It must lead them to ask the right questions, not just feed them the information they expect. That presupposes first that executives know what information they need. It demands further that they obtain that information on a regular basis. It finally requires that they systematically integrate the information into their decision making.

A few multinationals—Unilever, Coca-Cola, Nestlé, the Japanese trading companies, and a few big construction companies—have been working hard on building systems to gather and organize outside information. But in general, the majority of enterprises have yet to start the job.

Even big companies, in large part, will have to hire outsiders to help them. To think through what the business needs requires somebody who knows and understands the highly specialized information field. There is far too much information for any but specialists to find their way around. The sources are totally diverse. Companies can generate some of the information themselves, such as information about customers and noncustomers or about the technology in one's own field. But most of what enterprises need to know about the environment is obtainable only from outside sources—from all kinds of data banks and data services, from journals in many languages, from trade associations, from government publications, from World Bank reports and scientific papers, and from specialized studies.

Another reason why there is need for outside help is that the information has to be organized so as to question and challenge a company's strategy. To supply data is not enough. The data have to be integrated with strategy, they have to test a company's assumptions, and they must challenge a company's current outlook. One way to do that may be a new kind of software, information tailored to a specific group—say, to hospitals or to casualty insurance companies. The Lexis database supplies such information to lawyers, but it only gives answers; it does not ask questions. What we need are services that make specific suggestions about how to use the information, ask specific questions regarding the user's business and practices, and perhaps provide interactive consultation. Or we might "outsource" the outside-information system. Perhaps the most popular provider of the outside-information system, especially for smaller enterprises, will be the "inside outsider," the independent consultant.

Whichever way we satisfy it, the need for information on the environment where the major threats and opportunities are likely to arise will become increasingly urgent.

It may be argued that few of those information needs are new, and that is largely true. Conceptually, many of the new measurements have been discussed for many years and in many places. What is new is the technical data-processing ability. It enables us to do quickly and cheaply what, only a few short years ago, would have been laborious and very expensive. Seventy years ago the time and motion study made traditional cost accounting possible. Computers have now made activity-based cost accounting possible; without them, it would be practically impossible.

But that argument misses the point. What is important is not the tools. It is the concepts behind them. They convert what were always seen as discrete techniques to be used in isolation and for separate purposes into one integrated information system. That system then makes possible business diagnosis, business strategy, and business decisions. That is a new and radically different view of the meaning and purpose of information: as a measurement on which to base future action rather than as a postmortem and a record of what has already happened.

The command-and-control organization that first emerged in the 1870s might be compared to an organism held together by its shell. The corporation that is now emerging is being designed around a skeleton: *information,* both the corporation's new integrating system and its articulation.

Our traditional mind-set—even if we use sophisticated mathematical techniques and impenetrable sociological jargon—has always somehow perceived business as buying cheap and selling dear. The new approach defines a business as the organization that adds value and creates wealth.

8.

MANAGEMENT BY
OBJECTIVES AND
SELF-CONTROL

Any business enterprise must build a true team and weld individual efforts into a common effort. Each member of the enterprise contributes something different, but they must all contribute toward a common goal. Their efforts must all pull in the same direction, and their contributions must fit together to produce a whole—without gaps, without friction, without unnecessary duplication of effort.

Business performance therefore requires that each job be directed toward the objectives of the whole business. And in particular each manager's job must be focused on the success of the whole. The performance that is expected of the manager must be derived from the performance goals of the business; his results must be measured by the contribution they make to the success of the enterprise. The manager must know and understand what the business goals demand of him in terms of performance, and his superior must know what contribution to demand and expect of him—and must judge him accordingly. If these requirements are not met, managers are misdirected. Their efforts are wasted. Instead of teamwork, there is friction, frustration, and conflict.

Management by objectives requires major effort and special instruments. For in the business enterprise, managers are not automatically directed toward a common goal.

A favorite story at management meetings is that of the three stonecutters who were asked what they were doing. The first replied, "I am making a living." The second kept on hammering while he said, "I am doing the best job of stonecutting in the entire county." The third one looked up with a visionary gleam in his eyes and said, "I am building a cathedral."

The third man is, of course, the true "manager." The first man knows what he wants to get out of the work and manages to do so. He is likely to give a "fair day's work for a fair day's pay."

It is the second man who is a problem. Workmanship is essential; without it no business can flourish; in fact, an organization becomes demoralized if it does not demand of its members the most scrupulous workmanship they are capable of. But there is always a danger that the true workman, the true professional, will believe that he is accomplishing something when in effect he is just polishing stones or collecting footnotes. Workmanship must be encouraged in the business enterprise. But it must always be related to the needs of the whole.

The number of highly educated specialists working in the business enterprise is bound to increase tremendously. And so will the level of workmanship demanded of those specialists. The tendency to make the craft or function an end in itself will therefore be even more marked than it is today. But at the same time, the new technology will demand much closer coordination among specialists. And it will demand that functional men and women even at the lowest management level see the business as a whole and understand what it requires of them. The new technology will need both the drive for excellence in workmanship and the consistent direction of managers at all levels toward the common goal.

Misdirection

The hierarchical structure of management aggravates the danger. What the "boss" does and says, his most casual remarks, habits, even mannerisms, tend to appear to subordinates as calculated, planned, and meaningful.

"All you ever hear around the place is human-relations talk; but when the boss calls you on the carpet it is always because the burden figure is too high; and when it comes to promoting a guy, the plums always go to those who do the best job filling out accounting-department forms." This is one of the most common tunes, sung with infinite variations on every level of management. It leads to poor performance—even in cutting the burden figure. It also expresses loss of confidence in, and absence of respect for, the company and its management.

Yet the manager who so misdirects his subordinates does not intend to do so. He genuinely considers human relations to be the most important task of his plant managers. But he talks about the burden figure because he feels that he has to establish himself with his men as a "practical man," or because he thinks that he shows familiarity with their problems by talking "shop" with them. He stresses the accounting-department forms only because they annoy him as much as they do his men—or he may just not want to have any more trouble with the comptroller than he can help. But to his subordinates these reasons are hidden; all they see and hear is the question about the burden figure, the emphasis on forms.

The solution to this problem requires a structure of management that focuses both the manager's and his boss's eyes on what the job—rather than the boss—demands. To stress behavior and attitudes—as does a good deal of current management literature—cannot solve the problem. It is likely instead to aggravate it by making managers self-conscious in their relationships. Indeed, everyone familiar with business today has seen situations in which a manager's attempt to avoid misdirection through changing his behavior has converted a fairly satisfactory relationship into a nightmare of embarrassment and misunderstanding. The manager himself has become so self-conscious as to lose all easy relationship with his employees. And the employees in turn react with: "So help us, the old man has read a book; we used to know what he wanted of us, now we have to guess."

What Should the Objectives Be?

Each manager, from the "big boss" down to the production foreman or the chief clerk, needs clearly spelled-out objectives. Those objectives should lay out what performance the man's own managerial unit is supposed to produce. They should lay out what contribution he and his unit are expected to make to help other units obtain their objectives. Finally, they should spell out what contribution the manager can expect from other units toward the attainment of his own objectives. Right from the start, in other words, emphasis should be on teamwork and team results.

These objectives should always derive from the goals of the business enterprise. In one company, I have found it practicable and effective to provide even a foreman with a detailed statement of not only his own objectives but those of the company and of the manufacturing department. Even though the company is so large as to make the distance between the individual foreman's production and the company's total output all but astronomical, the result has been a significant increase in production. Indeed, this must follow if we mean it when we say that the foreman is "part of management." For it is the definition of a manager that in what he does he takes responsibility for the whole—that, in cutting stone, he "builds the cathedral."

The objectives of every manager should spell out his contribution to the attainment of company goals in *all areas* of the business. Obviously, not every manager has a direct contribution to make in every area. The contribution that marketing makes to productivity, for example, may be very small. But if a manager and his unit are not expected to contribute toward any one of the areas that significantly affect prosperity and survival of the business, this fact should be clearly brought out. For managers must understand that business results depend on a balance of efforts and output in a number of areas. This is necessary both to give full scope to the craftsmanship of each function and specialty, and to prevent the empire building and clannish jealousies of the various functions and specialties. It is necessary also to avoid overemphasis on any one key area.

To obtain balanced efforts, the objectives of all managers on all levels and in all areas should also be keyed to both short-range and long-range considerations. And, of course, all objectives should always contain both the tangible business objectives and the intangible objectives for manager organization and development, worker performance and attitude, and public responsibility. Anything else is shortsighted and impractical.

Management by "Drives"

Proper management requires balanced stress on objectives, especially by top management. It rules out the common and pernicious business malpractice: management by "crisis" and "drives."

There may be companies in which management people do not say, "The only way we ever get anything done around here is by making a drive on it." Yet "management by drive" is the rule rather than the exception. That things always collapse into the status quo ante three weeks after the drive is over, everybody knows and apparently expects. The only result of an "economy drive" is likely to be that messengers and typists get fired, and that $15,000 executives are forced to do $50-a-week work typing their own letters. And yet many managements have not drawn the obvious conclusion that drives are, after all, not the way to get things done.

But over and above its ineffectiveness, management by drive

misdirects. It puts all emphasis on one phase of the job to the inevitable detriment of everything else.

"For four weeks we cut inventories," a case-hardened veteran of management by crisis once summed it up. "Then we have four weeks of cost-cutting, followed by four weeks of human relations. We just have time to push customer service and courtesy for a month. And then the inventory is back where it was when we started. We don't even try to do our job. All management talks about, thinks about, preaches about, is last week's inventory figure or this week's customer complaints. How we do the rest of the job they don't even want to know."

In an organization that manages by drives, people either neglect their job to get on with the current drive, or silently organize for collective sabotage of the drive to get their work done. In either event they become deaf to the cry of "wolf." And when the real crisis comes, when all hands should drop everything and pitch in, they treat it as just another case of management-created hysteria.

Management by drive, like management by "bellows and meat ax," is a sure sign of confusion. It is an admission of incompetence. It is a sign that management does not know how to plan. But, above all, it is a sign that the company does not know what to expect of its managers—that, not knowing how to direct them, it misdirects them.

How Should Objectives Be Set and by Whom?

By definition, a manager is responsible for the contribution that his or her component makes to the larger unit above and eventually to the enterprise. The manager's performance aims upward rather than downward. This means that the goals of each manager's job must be defined by the contribution he has to make to the success of the larger unit of which he is a part. The objectives of the district sales manager's job should be defined by the contribution he and his district sales force have to make to the sales department; the objectives of the project engineer's job, by the contribution he, his engineers and draftsmen make to the engineering department. The objectives of the general manager of a decentralized division should be defined by the contribution his division has to make to the objectives of the parent company.

This requires each manager to develop and set the objectives of his unit himself. Higher management must, of course, reserve the power to approve or disapprove those objectives. But their development is part of a manager's responsibility; indeed, it is his first responsibility. It means, too, that every manager should responsibly participate in the development of the objectives of the higher unit of which his is a part. To "give him a sense of participation" (to use a pet phrase of the "human relations" jargon) is not enough. Being a manager demands the assumption of a genuine responsibility. Precisely because his aims should reflect the objective needs of the business, rather than merely what the individual manager wants, he must commit himself to them with a positive act of assent. He must know and understand the ultimate business goals, what is expected of him and why, what he will be measured against and how. There must be a "meeting of minds" within the entire management of each unit. This can be achieved only when each of the contributing managers is expected to think through what the unit objectives are, is led, in others words, to participate actively and responsibly in the work of defining them. And only if his lower managers participate in this way can the higher manager know what to expect of them and can make exacting demands.

This is so important that some of the most effective managers I know go one step further. They have each of their subordinates write a "manager's letter" twice a year. In this letter to his superior, each manager first defines the objectives of his superior's job and of his own job as he sees them. He then sets down the performance standards that he believes are being applied to him. Next, he lists the things he must do himself to attain these goals—and what he considers the major obstacles within his own unit. He lists the things his superior and the company do that help him and the things that hamper him. Finally, he outlines what he proposes to do during the next year to reach his goals. If his superior accepts this statement, the "manager's letter" becomes the charter under which the manager operates.

This device, like no other I have seen, brings out how easily the unconsidered and casual remarks of even the best "boss" can confuse and misdirect. One large company has used the "manager's letter" for ten years. Yet almost every letter still lists as objectives and standards things that completely baffle the superior to whom the letter is addressed. And whenever he asks, "What is this?" he gets the answer, "Don't you remember what you said last spring going down with me in the elevator?"

The "manager's letter" also brings out whatever inconsistencies there are in the demands made on a man by his superior and by the company. Does the superior demand both speed and high quality when he can get only one or the other? And what compromise is needed in the interest of the company? Does he demand initiative and judgment of his men but also that they check back with him before they do anything? Does he ask for their ideas and suggestions but never use them or discuss them? Does the company expect a small engineering force to be available immediately whenever something goes wrong in the plant, and yet bend all its efforts to the completion of new designs? Does it expect a manager to maintain high standards of performance but forbid him to remove poor performers? Does it create the conditions under which people say, I can get the work done as long as I can keep the boss from knowing what I am doing?

These are common situations. They undermine spirit and performance. The "manager's letter" may not prevent them. But at least it brings them out in the open, shows where compromises have to be made, objectives have to be thought through, priorities have to be established, behavior has to be changed.

As this device illustrates: managing managers requires special efforts not only to establish common direction, but to eliminate misdirection. Mutual understanding can never be attained by "communications down," can never be created by talking. It can result only from "communications up." It requires both the superior's willingness to listen and a tool especially designed to make lower managers heard.

Self-control through Measurements

The greatest advantage of management by objectives is perhaps that it makes it possible for a manager to control his own performance. Self-control means stronger motivation: a desire to do the best rather than just enough to get by. It means higher performance goals and broader vision. Even if management by objectives was not necessary to give the enterprise the unity of direction and effort of a management team, it would be necessary to make possible management by self-control.

Indeed, one of the major contributions of management by objectives is that it enables us to substitute management by self-control for management by domination.

That management by self-control is highly desirable will hardly be disputed in America or in American business today. Its acceptance underlies all the talk of "pushing decisions down to the lowest possible level," or of "paying people for results." But to make management by self-control a reality requires more than acceptance of the concept as right and desirable. It requires new tools and far-reaching changes in traditional thinking and practices.

To be able to control his own performance, a manager needs to know more than what his goals are. He must be able to measure his performance and results against the goal. It should indeed be an invariable practice to supply managers with clear and common measurements in all key areas of a business. Those measurements need not be rigidly quantitative; nor need they be exact. But they have to be clear, simple, and rational. They have to be relevant and direct attention and efforts where they should go. They have to be reliable—at least to the point where their margin of error is acknowledged and understood. And they have to be, so to speak, self-announcing, understandable without complicated interpretation or philosophical discussion.

Each manager should have the information he needs to measure his own performance and should receive it soon enough to make any changes necessary for the desired results. And this information should go to the manager himself, and not to his superior. It should be the means of self-control, not a tool of control from above.

This needs particular stress today, when our ability to obtain such information is growing rapidly as a result of technological progress in information gathering, analysis, and synthesis. Up till now information on important facts was either not obtainable at all, or could be assembled only so late as to be of little but historical interest. This former inability to produce measuring information was not an unmixed curse. For while it made effective self-control difficult, it also made difficult effective control of a manager from above; in the absence of information with which to control him, the manager had to be allowed to work as he saw fit.

Our new ability to produce measuring information will make possible effective self-control; and if so used, it will lead to a tremendous advance in the effectiveness and performance of management. But if this new ability is abused in order to impose control on managers from above, the new technology will inflict incalculable harm by demoralizing management, and by seriously lowering the effectiveness of managers.

That information can be effectively used for self-control is shown by the example of General Electric.

General Electric has a special control service—the traveling auditors. The auditors study every one of the managerial units of the company thoroughly at least once a year. But their report goes to the manager of the unit studied. There can be little doubt that the feeling of confidence and trust in the company that even casual contact with General Electric managers reveals is directly traceable to this practice of using information for self-control rather than for control from above.

But the General Electric practice is by no means common or generally understood. Typical management thinking is much closer to the practice exemplified by a large chemical company.

In this company a control section audits every one of the managerial units of the company. The results of the audits do not go, however, to the managers audited. They go only to the president, who then calls in the managers to confront them with the audit of their operations. What this has done to morale is shown in the nickname the company's managers have given the control section: "the president's Gestapo." Indeed, more and more managers are now running their units not to obtain the best performance but to obtain the best showing on the control-section audits.

This should not be misunderstood as advocacy of low performance standards or absence of control. On the contrary, management by objectives and self-control is primarily a means to achieve standards higher than are to be found in most companies today. And every manager should be held strictly accountable for the results of his performance.

But what he does to reach those results he—and only he—should control. It should be clearly understood what behavior and methods the company bars as unethical, unprofessional, or unsound. But within these limits, every manager must be free to decide what he has to do. And only if he has all the information regarding his operations can he fully be held accountable for results.

The Proper Use of Reports and Procedures

Management by self-control requires complete rethinking concerning our use of reports, procedures, and forms.

Reports and procedures are necessary tools. But few tools can be so easily misused, and few can do as much damage. For reports and procedures, when misused, cease to be tools and become malignant masters.

There are three common misuses of reports and procedures.

The first is the all too common belief that procedures are instruments of morality. They are not; their principle is exclusively that of economy. They never decide what should be done, only how it might be done most expeditiously. Problems of right conduct can never be "proceduralized" (surely the most horrible word in the bureaucrat's jargon); conversely, right conduct can never be established by procedure.

The second misuse is to consider procedures a substitute for judgment. Procedures can work only where judgment is no longer required, that is, in the repetitive situation for whose handling the judgment has already been supplied and tested. Our civilization suffers from a superstitious belief in the magical effect of printed forms. And the superstition is most dangerous when it leads us into trying to handle the exceptional, nonroutine situation by procedure. In fact, it is the test of a good procedure that it quickly identifies the situations that, even in the most routine of processes, do not fit the pattern but require special handling and decision based on judgment.

But the most common misuse of reports and procedures is as an instrument of control from above. This is particularly true of those that aim at supplying information to higher management—the "forms" of everyday business life. The common case of the plant manager who has to fill out twenty forms to supply accountants, engineers, or staff people in the central office with information he himself does not need, is only one of thousands of examples. As a result, the man's attention is directed away from his own job. The things he is asked about or required to do for control purposes come to appear to him as reflections of what the company wants of him, become to him the essence of his job; while resenting them, he tends to put effort into these things rather than into his own job. Eventually, his boss, too, is misdirected, if not hypnotized, by the procedure.

A large insurance company, a few years ago, started a big program for the "improvement of management." To this end it built up a strong central-office organization concerned with such things as renewal ratios, claims settlement, selling costs, sales methods, etc. This organization did excellent work—top management learned a lot about running an insurance company. But actual performance has been going down ever since. For the managers in the field spend more and more time filling out reports, less and less doing their work. Worse still, they soon learned to subordinate performance to a "good showing." Not only did performance go to pieces—morale suffered even more. Top management and its staff experts came to be viewed by the field managers as enemies to be outsmarted or at least kept as far away as possible.

Similar stories exist ad infinitum—in every industry and in companies of every size. To some extent the situation is caused by the fallacy of the "staff" concept. But, above all, it is the result of the misuse of procedures as control.

Reports and procedures should be kept to a minimum, and used only when they save time and labor. They should be as simple as possible.

One of our leading company presidents tells the following story on himself. Fifteen years ago he bought for his company a small independent plant in Los Angeles. The plant had been making a profit of $250,000 a year; and it was purchased on that basis. When going through the plant with the owner—who stayed on as plant manager—the president asked, "How do you determine your pricing?" "That's easy," the former owner answered; "we just quote ten cents per thousand less than your company does." "And how do you control your costs?" was the next question. "That's easy," was the answer; "we know what we pay for raw materials and labor and what production we ought to get for the money." "And how do you control your overhead?" was the final question. "We don't bother about it."

Well, thought the president, we can certainly save a lot of money here by introducing our thorough controls. But a year later the profit of the plant was down to $125,000; sales had remained the same and prices had remained the same; but the introduction of complex procedures had eaten up half the profit.

Every business should regularly find out whether it needs all the reports and procedures it uses. At least once every five years, every form should be put on trial for its life. I once had to recommend an even more drastic measure to clear up a situation in which reports and forms, luxuriating like the Amazon rain forest, threatened to choke the life out of an old established utility company. I suggested that all reports be suspended simultaneously for two months, and only those be allowed to return that managers still demanded after living without them. This cut reports and forms in the company by three quarters.

Reports and procedures should focus only on the performance needed to achieve results in the key areas. To "control" everything is to control nothing. And to attempt to control the irrelevant always misdirects.

Finally, reports and procedures should be the tool of the man who fills them out. They must never themselves become the measure of his performance. A man must never be judged by the quality of the production forms he fills out—unless he be the clerk in charge of those forms. He must always be judged by his production performance. And the only way to make sure of this is by having him fill out no forms, make no reports, except those he needs himself to achieve performance.

A Philosophy of Management

What the business enterprise needs is a principle of management that will give full scope to individual strength and responsibility, and at the same time give common direction of vision and effort, establish team work, and harmonize the goals of the individual with the commonweal.

The only principle that can do this is management by objectives and self-control. It makes the commonweal the aim of every manager. It substitutes for control from outside the stricter, more exacting and more effective control from the inside. It motivates the manager to action not because somebody tells him to do something or talks him into doing it, but because the objective needs of his task demand it. He acts not because somebody wants him to but because he himself decides that he has to—he acts, in other words, as a free man.

The word "philosophy" is tossed around with happy abandon these days in management circles. I have even seen a dissertation, signed by a vice president, on the "philosophy of handling purchase requisitions" (as far as I could figure out, "philosophy" here meant that purchase requisitions had to be in triplicate). But management by objectives and self-control may legitimately be called a "philosophy" of management. It rests on a concept of the job of management. It rests on an analysis of the specific needs of the management group and the obstacles it faces. It rests on a concept of human action, human behavior, and human motivation. Finally, it applies to every manager, whatever his level and function, and to any business enterprise whether large or small. It ensures performance by converting objective needs into personal goals. And this is genuine freedom, freedom under the law.

9.

PICKING PEOPLE—
THE BASIC RULES

Executives spend more time on managing people and making people decisions than on anything else, and they should. No other decisions are so long-lasting in their consequences or so difficult to unmake. And yet, by and large, executives make poor promotion and staffing decisions. By all accounts, their batting average is no better than .333: at most one-third of such decisions turn out right; one-third are minimally effective; and one-third are outright failures.

In no other area of management would we put up with such miserable performance. Indeed, we need not and should not. Managers making people decisions will never be perfect, of course. But they should come pretty close to batting 1.000, especially because in no other area of management do we know so much.

Some executives' people decisions have, however, approached perfection. At the time of Pearl Harbor, every single general officer in the U.S. Army was overage. Although none of the younger men had been tested in combat or in a significant troop command, the United States came out of World War II with the largest corps of competent general officers any army has ever had. General George C. Marshall, the army's chief of staff, had personally chosen each man. Not all were great successes, but practically none was an outright failure.

In the forty or so years during which he ran General Motors, Alfred P. Sloan Jr. picked every GM executive—down to the manufacturing managers, controllers, engineering managers, and master mechanics at even the smallest accessory division. By today's standards, Sloan's vision and values may seem narrow. They were. He was concerned only with performance in and for GM. Nonetheless, his long-term performance in placing people in the right jobs was flawless.

The Basic Principles

There is no such thing as an infallible judge of people, at least not on this side of the Pearly Gates. There are, however, a few executives who take their people decisions seriously and work at them.

Marshall and Sloan were about as different as two human beings can be, but they followed, and quite consciously, much the same principles in making people decisions.

1. If I put a person into a job and he or she does not perform, I have made a mistake. I have no business blaming that person, no business invoking the "Peter Principle," no business complaining. I have made a mistake.

2. The soldier has a right to competent command, was already an old maxim at the time of Julius Caesar. It is the duty of managers to make sure that the responsible people in their organizations perform.

3. Of all the decisions an executive makes, none is as important as the decisions about people because they determine the performance capacity of the organization. Therefore, I'd better make these decisions well.

4. The one "don't": Don't give new people major assignments, for doing so only compounds the risks. Give this sort of assignment

to someone whose behavior and habits you know and who has earned trust and credibility within your organization. Put a high-level newcomer first into an established position where the expectations are known and help is available.

The Decision Steps

Just as there are only a few basic principles, there are only a few important steps to follow in making effective promotion and staffing decisions.

1. *Think through the assignment.* Job descriptions may last a long time. In one large manufacturing company, for example, the job description for the position of division general manager has hardly changed since the company began to decentralize thirty years ago. Indeed, the job description for bishops in the Roman Catholic Church has not changed at all since canon law was first codified in the thirteenth century. But assignments change all the time, and unpredictably.

Once in the early 1940s, I mentioned to Alfred Sloan that he seemed to me to spend an inordinate amount of time pondering the assignment of a fairly low-level job—general sales manager of a small accessory division—before choosing among three equally qualified candidates. "Look at the assignment the last few times we had to fill the same job," Sloan answered. To my surprise, I found that the terms of the assignment were quite different on each occasion.

When putting a man in as division commander during World War II, George Marshall always looked first at the nature of the assignment for the next eighteen months or two years. To raise a division and train it is one assignment. To lead it in combat is quite another. To take command of a division that has been badly mauled and restore its morale and fighting strength is another still.

When the task is to select a new regional sales manager, the responsible executive must first know what the heart of the assignment is: to recruit and train new salespeople because, say, the present sales force is nearing retirement age? Or is it to open up new markets because the company's products, though doing well with old-line industries in the region, have not been able to penetrate new and growing markets? Or, because the bulk of sales still comes from products that are twenty-five years old, is it to establish a market presence for the company's new products? Each of these is a different assignment and requires a different kind of person.

2. *Look at a number of potentially qualified people.* The controlling word here is *number.* Formal qualifications are a minimum for consideration; their absence disqualifies the candidate automatically. Equally important, the person and the assignment need to fit each other. To make an effective decision, an executive should look at three to five qualified candidates.

3. *Think hard about how to look at these candidates.* If an executive has studied the assignment, he or she understands what a new person would need to do with high priority and concentrated effort. The central question is not, What can this or that candidate do or not do? It is, rather, What are the strengths each possesses and are these the right strengths for the assignment? Weaknesses are limitations, which may, of course, rule a candidate out. For instance, a person may be excellently qualified for the technical aspects of a job, but if the assignment requires above all the ability to build a team and this ability is lacking, then the fit is not right.

But effective executives do not start out by looking at weaknesses. You cannot build performance on weaknesses. You can build only on strengths.

Both Marshall and Sloan were highly demanding men, but both knew that what matters is the ability to do the assignment. If that exists, the company can always supply the rest. If it does not exist, the rest is useless.

If, for instance, a division needed an officer for a training assignment, Marshall looked for people who could turn recruits into soldiers. Usually every man who was good at this task had serious weaknesses in other areas. One was not particularly effective as a tactical commander and was positively hopeless when it came to strategy. Another had foot-in-mouth disease and got into trouble with the press. A third was vain, arrogant, egotistical, and fought constantly with his commanding officer. Never mind, could he train recruits? If the answer was yes—and especially if the answer was "he's the best"—he got the job.

In picking members of their cabinets, Franklin Roosevelt and Harry Truman said, in effect, Never mind personal weaknesses. Tell me first what each of them can do. It may not be coincidence that these two Presidents had the strongest Cabinets in twentieth-century U.S. history.

4. *Discuss each of the candidates with several people who have worked with them.* One executive's judgment alone is worthless. Because all of us have first impressions, prejudices, likes, and dislikes, we need to listen to what other people think. When the military picks general officers or the Catholic Church picks bishops, this kind of extensive discussion is a formal step in their selection process. Competent executives do it informally. Hermann Abs, the former head of Deutsche Bank, picked more successful chief executives in recent times than anyone else. He personally chose most of the top-level managers who pulled off the postwar German "economic miracle," and he checked out each of them first with three or four of their former bosses or colleagues.

5. *Make sure the appointee understands the job.* After the appointee has been in a new job for three or four months, he or she should be focusing on the demands of that job rather than on the requirements of preceding assignments. It is the executive's responsibility to call that person in and say, "You have now been regional sales manager [or whatever] for three months. What do you have to do to be a success in your new job? Think it through and come back in a week or ten days and show me in writing. But I can tell you one thing right away: the things you did to get the promotion are almost certainly the wrong things to do now."

If you do not follow this step, don't blame the candidate for poor performance. Blame yourself. You have failed in your duty as a manager.

The largest single source of failed promotions—and I know of no greater waste in U.S. management—is the failure to think through, and help others think through, what a new job requires.

All too typical is the brilliant former student of mine who telephoned a few months ago, almost in tears. "I got my first big chance a year ago," he said. "My company made me engineering manager. Now they tell me that I'm through. And yet I've done a better job than ever before. I have actually designed three successful new products for which we'll get patents."

It is only human to say to ourselves, I must have done something right or I would not have gotten the big new job. Therefore, I had better do more of what I did to get the promotion now that I have it. It is not intuitively obvious to most people that a new and different job requires new and different behavior. Almost fifty years ago, a boss of mine challenged me four months after he had advanced me to a far more responsible position. Until he called me in, I had continued to do what I had done before. To his credit, he understood that it was his responsibility to make me see that a new job means different behavior, a different focus, and different relationships.

The High-Risk Decisions

Even if executives follow all these steps, some of their people decisions will still fail. These are, for the most part, the high-risk decisions that nevertheless have to be taken.

There is, for example, high risk in picking managers in professional organizations—in a research lab, say, or an engineering or corporate legal department. Professionals do not readily accept as their boss someone whose credentials in the field they do not respect. In choosing a manager of engineering, the choices are therefore limited to the top-flight engineers in the department. Yet there is no correlation (unless it be a negative one) between performance as a bench engineer and performance as a manager. Much the same is true when a high-performing operating manager gets a promotion to a staff job at headquarters or a staff expert moves into a line position. Temperamentally, operating people are frequently unsuited to the tensions, frustrations, and relationships of staff work, and vice versa. The first-rate regional sales manager may well become totally ineffective if promoted into market research, sales forecasting, or pricing.

We do not know how to test or predict whether a person's temperament will be suited to a new environment. We can find this out only by experience. If a move from one kind of work to another does not pan out, the executive who made the decision has to remove the misfit, and fast. But that executive also has to say, I made a mistake, and it is my job to correct it. To keep misfits in a job they cannot do is not being kind; it is being cruel. But there is also no reason to let the person go. A company can always use a good bench engineer, a good analyst, a good sales manager. The proper course of action—and it works most times—is to offer the misfit a return to the old job or an equivalent.

People decisions may also fail because a job has become what New England ship captains 150 years ago called a "widow-maker." When a clipper ship, no matter how well designed and constructed, began to have fatal "accidents," the owners did not redesign or rebuild the ship. They broke it up as fast as possible.

Widow-makers—that is, jobs that regularly defeat even good people—appear most often when a company grows or changes fast. For instance, in the 1960s and early 1970s, the job of "international vice president" in U.S. banks became a widow-maker. It had always been an easy job to fill. In fact, it had long been considered a job into which banks could safely put also-rans and expect them to perform well. Then, suddenly, the job began to defeat one new appointee after another. What had happened, as hindsight now tells us, was that international activity quickly and without warning had become an integral part of the daily business of major banks and their corporate customers. What had been until then an easy job became, literally, a "nonjob" that nobody could do.

Whenever a job defeats two people in a row, who in their earlier assignments had performed well, a company has a widow-maker on its hands. When this happens, a responsible executive should not ask the headhunter for a universal genius. Instead abolish the job. Any job that ordinarily competent people cannot perform is a job that cannot be staffed. Unless changed, it will predictably defeat the third appointee the way it defeated the first two.

Making the right people decisions is the ultimate means of controlling an organization well. Such decisions reveal how competent management is, what its values are, and whether it takes its job seriously. No matter how hard managers try to keep their decisions a secret—and some still try hard—people decisions cannot be hidden. They are eminently visible.

Executives often cannot judge whether a strategic move is a wise one. Nor are they necessarily interested. "I don't know why we are buying this business in Australia, but it won't interfere with what we are doing here in Fort Worth" is a common reaction. But when the same executives read that "Joe Smith has been made controller in the XYZ division," they usually know Joe much better than top management does. These executives should be able to say, "Joe deserves the promotion; he is an excellent choice, just the person that division needs to get the controls appropriate for its rapid growth."

If, however, Joe got promoted because he is a politician, everybody will know it. They will all say to themselves, Okay, that is the way to get ahead in this company. They will despise their management for forcing them to become politicians but will either quit or become politicians themselves in the end. As we have known for a long time, people in organizations tend to be influenced by the ways they see others being rewarded. And when the rewards go to nonperformance, to flattery, or to mere cleverness, the organization will soon decline into nonperformance, flattery, or cleverness.

Executives who do not make the effort to get their people decisions right do more than risk poor performance. They risk their organization's respect.

10.

THE ENTREPRENEURIAL
BUSINESS

B ig businesses don't innovate, says the conventional wisdom. This sounds plausible enough. True, the new, major innovations of this century did not come out of the old, large businesses of their time. The railroads did not spawn the automobile or the truck; they did not even try. And though the automobile companies did try (Ford and General Motors both pioneered in aviation and aerospace), all of today's large aircraft and aviation companies have evolved out of separate new ventures. Similarly, today's giants of the pharmaceutical industry are, in the main, companies that were small or nonexistent fifty years ago when the first modern drugs were developed. Every one of the giants of the electronics industry—General Electric, Westinghouse, and RCA in the United States; Siemens and Philips on the Continent; Toshiba in Japan—rushed into computers in the 1950s. Not one was successful. The field is dominated by IBM, a company that was barely middle-sized and most definitely not high-tech forty years ago.

And yet the all but universal belief that large businesses do not and cannot innovate is not even a half-truth; rather, it is a misunderstanding.

In the first place, there are plenty of exceptions, plenty of large companies that have done well as entrepreneurs and innovators. In the United States, there is Johnson & Johnson in hygiene and health care, and 3M in highly engineered products for both industrial and consumer markets. Citibank, America's and the world's largest nongovernmental financial institution, well over a century old, has been a major innovator in many areas of banking and finance. In Germany, Hoechst—one of the world's largest chemical companies, and more than 125 years old by now—has become a successful innovator in the pharmaceutical industry.

Second, it is not true that "bigness" is an obstacle to entrepreneurship and innovation. In discussions of entrepreneurship, one hears a great deal about the "bureaucracy" of big organizations and of their "conservatism." Both exist, of course, and they are serious impediments to entrepreneurship and innovation—but to all other performance just as much. And yet the record shows unambiguously that among existing enterprises, whether business or public-sector institutions, the small ones are least entrepreneurial and least innovative. Among existing entrepreneurial businesses there are a great many very big ones; the list above could have been enlarged without difficulty to one hundred companies from all over the world, and a list of innovative public-service institutions would also include a good many large ones.

It is not size that is an impediment to entrepreneurship and innovation; it is the existing operation itself, and especially the existing *successful* operation. And it is easier for a big or at least a fair-sized company to surmount this obstacle than it is for a small one. Operating anything—a manufacturing plant, a technology, a product line, a distribution system—requires constant effort and unremitting attention. The one thing that can be guaranteed in any kind of operation is the daily crisis. The daily crisis cannot be postponed; it has to be dealt with right away. And the existing operation demands high priority and deserves it. The new always looks so small, so puny, so unpromising next to the size and performance of maturity.

Where the conventional wisdom goes wrong is in its assumption that entrepreneurship and innovation are natural, creative, or spontaneous. If entrepreneurship and innovation do not well up in an organization, something must be stifling them. That only a minority of existing successful businesses are entrepreneurial and innovative is thus seen as conclusive evidence that existing businesses quench the entrepreneurial spirit. But entrepreneurship is not "natural"; it is not "creative." It is work. Hence, the correct conclusion from the evidence is the opposite of the one commonly reached. That a substantial number of existing businesses, and among them a goodly number of fair-sized, big, and very big ones, succeed as entrepreneurs and innovators indicates that entrepreneurship and innovation can be achieved by any business. But they must be consciously striven for. They can be learned, but it requires effort. Entrepreneurial businesses treat entrepreneurship as a duty. They are disciplined about it . . . they work at it . . . they practice it.

Structures

People work within a structure.

For the existing business to be capable of innovation, it has to create a structure that allows people to be entrepreneurial. It has to devise relationships that center on entrepreneurship. It has to make sure that its incentives, its compensation, personnel decisions, and policies, all reward the right entrepreneurial behavior and do not penalize it.

1. This means, first, that the entrepreneurial, the new, has to be organized separately from the old and existing. Whenever we have tried to make an existing unit the carrier of the entrepreneurial project, we have failed.

One reason is that the existing business always requires time and effort on the part of the people responsible for it, and deserves the priority they give it. The new always looks so puny—so unpromising—next to the reality of the massive, ongoing business. The existing business, after all, has to nourish the struggling innovation. But the "crisis" in today's business has to be attended to as well. The people responsible for an existing business will therefore always be tempted to postpone action on anything new, entrepreneurial, or innovative until it is too late. No matter what has been tried—and we have now been trying every conceivable mechanism for thirty or forty years—existing units have been found to be capable mainly of extending, modifying, and adapting what already is in existence. The new belongs elsewhere.

2. This means also that there has to be a special locus for the new venture within the organization, and it has to be pretty high up. Even though the new project, by virtue of its current size, revenues, and markets, does not rank with existing products, somebody in top management must have the specific assignment to work on tomorrow as an entrepreneur and innovator.

This need not be a full-time job; in the smaller business, it very often cannot be a full-time job. But it needs to be a clearly defined job and one for which somebody with authority and prestige is fully accountable.

The new project is an infant and will remain one for the foreseeable future, and infants belong in the nursery. The "adults," that is, the executives in charge of existing businesses or products, will have neither time nor understanding for the infant project. They cannot afford to be bothered.

Disregard of this rule cost a major machine-tool manufacturer its leadership in robotics.

The company had the basic patents on machine tools for automated mass production. It had excellent engineering, an excellent reputation, and first-rate manufacturing. Everyone in the early years of factory automation—around 1975—expected it to emerge as the leader. Ten years later it had dropped out of the race entirely. The company had placed the unit charged with the development of machine tools for automated production three or four levels down in the organization, and had it report to people charged with designing, making, and selling the company's traditional machine-tool lines. Those people were supportive; in fact, the work on robotics had been mainly their idea. But they were far too busy defending their traditional lines against a lot of new competitors such as the Japanese, redesigning them to fit new specifications, demonstrating, marketing, financing, and servicing them. Whenever the people in charge of the "infant" went to their bosses for a decision, they were told, "I have no time now, come back next week." Robotics were, after all, only a promise; the existing machine-tool lines produced millions of dollars each year.

Unfortunately, this is a common error.

The best, and perhaps the only, way to avoid killing off the new by sheer neglect is to set up the innovative project from the start as a separate business.

The best-known practitioners of this approach are three American companies: Procter & Gamble—the soap, detergent, edible oil, and food producer—a very large and aggressively entrepreneurial company; Johnson & Johnson, the hygiene and health-care supplier; and 3M, a major manufacturer of industrial and consumer products. These three companies differ in the details of practice but essentially all three have the same policy. They set up the new venture as a separate business from the beginning and put a project manager in charge. The project manager remains in charge until the project is either abandoned or has achieved its objective and become a full-fledged business. And until then, the project manager can mobilize all the skills as they are needed—research, manufacturing, finance, marketing—and put them to work on the project team.

3. There is another reason why a new, innovative effort is best set up separately: to keep away from it the burdens it cannot yet carry. Both the investment in a new product line and its returns should, for instance, not be included in the traditional return-on-investment analysis until the product line has been on the market for a number of years. To ask the fledgling development to shoulder the full burdens an existing business imposes on its units is like asking a six-year-old to go on a long hike carrying a sixty-pound pack;

neither will get very far. And yet the existing business has requirements with respect to accounting, to personnel policy, to reporting of all kinds, which it cannot easily waive.

The innovative effort and the unit that carries it require different policies, rules, and measurements in many areas.

I learned this many years ago in a major chemical company. Everybody knew that one of its central divisions had to produce new materials to stay in business. The plans for these materials were there, the scientific work had been done . . . but nothing happened. Every year there was another excuse. Finally, the division's general manager spoke up at a review meeting: "My management group and I are compensated primarily on the basis of return on investment. The moment we spend money on developing the new materials, our return will go down by half for at least four years. Even if I am still here in four years time when we should show the first returns on these investments—and I doubt that the company will put up with me that long if profits are that much lower—I'm taking bread out of the mouths of all my associates in the meantime. Is it reasonable to expect us to do this?" The formula was changed and the developmental expenses for the new project were taken out of the return-on-investment figures. Within eighteen months the new materials were on the market. Two years later they had given the division leadership in its field, which it has retained to this day. Four years later the division doubled its profits.

The Don'ts

There are some things the entrepreneurial management of an existing business should not do.

1. The most important caveat is not to mix managerial units and entrepreneurial ones. Do not ever put the entrepreneurial into the existing managerial component. Do not make innovation an objective for people charged with running, exploiting, optimizing what already exists.

But it is also inadvisable—in fact, almost a guarantee of failure—for a business to try to become entrepreneurial without changing its basic policies and practices. To be an entrepreneur on the side rarely works.

In the last ten or fifteen years a great many large American companies have tried to go into joint ventures with entrepreneurs. Not one of these attempts has succeeded; the entrepreneurs found themselves stymied by policies, by basic rules, by a "climate" they felt was bureaucratic, stodgy, reactionary. But at the same time their partners, the people from the big company, could not figure out what the entrepreneurs were trying to do and thought them undisciplined, wild, visionary.

By and large, big companies have been successful as entrepreneurs only if they use their own people to build the venture. They have been successful only when they use people whom they understand and who understand them, people whom they trust and who in turn know how to get things done in the existing business; people, in other words, with whom one can work as partners. But this presupposes that the entire company is imbued with the entrepreneurial spirit, that it wants innovation and is reaching out for it, considering it both a necessity and an opportunity. It presupposes that the entire organization has been made "greedy for new things."

2. Innovative efforts that take the existing business out of its own field are rarely successful. Innovation had better not be "diversification." Whatever the benefits of diversification, it does not mix with entrepreneurship and innovation. The new is always sufficiently difficult not to attempt it in an area one does not understand. An existing business innovates where it has expertise, whether knowledge of market or knowledge of technology. Anything new will predictably get into trouble, and then one has to know the business. Diversification itself rarely works unless it, too, is built on commonality with the existing business, whether commonality of the market or commonality of the technology. Even then, as I have discussed elsewhere, diversification has its problems.

But if one adds to the difficulties and demands of diversification the difficulties and demands of entrepreneurship, the result is predictable disaster. So one innovates only where one understands.

3. Finally, it is almost always futile to avoid making one's own business entrepreneurial by "buying in," that is, by acquiring small entrepreneurial ventures. Acquisitions rarely work unless the company that does the acquiring is willing and able within a fairly short time to furnish management to the acquisition. The managers that have come with the acquired company rarely stay around very long. If they were owners, they have now become wealthy; if they were professional managers, they are likely to stay around only if given much bigger opportunities in the new, acquiring company. So, within a year or two, the acquirer has to furnish management to run the business that has been bought. This is particularly true when a nonentrepreneurial company buys an entrepreneurial one. The management people in the new acquired venture soon find that they cannot work with the people in their new parent company, and vice versa. I myself know of no case where "buying in" has worked.

A business that wants to be able to innovate, wants to have a chance to succeed and prosper in a time of rapid change, has to build entrepreneurial management into its own system. It has to adopt policies that create throughout the entire organization the desire to innovate and the habits of entrepreneurship and innovation. To be a successful entrepreneur, the existing business, large or small, has to be managed as an entrepreneurial business.

11.

THE NEW VENTURE

For the existing enterprise, whether business or public-service institution, the controlling word in the term "entrepreneurial management" is "entrepreneurial." For the new venture, it is "management." In the existing business, it is the existing that is the main obstacle to entrepreneurship. In the new venture, it is its absence.

The new venture has an idea. It may have a product or a service. It may even have sales, and sometimes quite a substantial volume of them. It surely has costs. And it may have revenues and even profits. What it does not have is a "business," a viable, operating, organized "present" in which people know where they are going, what they are supposed to do, and what the results are or should be. But unless a new venture develops into a new business and makes sure of being "managed," it will not survive no matter how brilliant the entrepreneurial idea, how much money it attracts, how good its products, or even how great the demand for them.

Refusal to accept these facts destroyed every single venture started by the nineteenth century's greatest inventor, Thomas Edison. Edison's ambition was to be a successful businessman and the head of a big company. He should have succeeded, for he was a superb business planner. He knew exactly how an electric power company had to be set up to exploit his invention of the light bulb. He knew exactly how to get all the money he could possibly need for his ventures. His products were immediate successes and the demand for them practically insatiable. But Edison remained an entrepreneur; or rather, he thought that "managing" meant being the boss. He refused to build a management team. And so every one of his four or five companies collapsed ignominiously once it got to middle size, and was saved only by booting Edison himself out and replacing him with professional management.

Entrepreneurial management in the new venture has four requirements:

It requires, first, a focus on the market.

It requires, second, financial foresight, and especially planning for cash flow and capital needs ahead.

It requires, third, building a top management team long before the new venture actually needs one and long before it can actually afford one.

And finally, it requires of the founding entrepreneur a decision in respect to his or her own role, area of work, and relationships.

The Need for Market Focus

A common explanation for the failure of a new venture to live up to its promise or even to survive at all is: "We were doing fine until these other people came and took our market away from us. We don't really understand it. What they offered wasn't so very different from what we had." Or one hears: "We were doing all right, but these other people started selling to customers we'd never even heard of and all of a sudden they had the market."

When a new venture does succeed, more often than not it is in a market other than the one it was originally intended to serve, with products or services not quite those with which it had set out, bought in large part by customers it did not even think of when it started, and used for a host of purposes besides the ones for which the products were first designed. If a new venture does not anticipate this, organizing itself to take advantage of the unexpected and unseen markets; if it is not totally market-focused, if not market-driven, then it will succeed only in creating an opportunity for a competitor.

A German chemist developed Novocain as the first local anesthetic in 1905. But he could not get the doctors to use it; they preferred total anesthesia (they only accepted Novocain during World War I). But totally unexpectedly, dentists began to use the stuff. Whereupon—or so the story goes—the chemist began to travel up and down Germany making speeches against Novocain's use in dentistry. He had not designed it for that purpose!

That reaction was somewhat extreme, I admit. Still, entrepreneurs *know* what their innovation is meant to do. And if some other use for it appears, they tend to resent it. They may not actually refuse to serve customers they have not "planned" for, but they are likely to make it clear that those customers are not welcome.

This is what happened with the computer. The company that had the first computer, Univac, knew that its magnificent machine was designed for scientific work. And so it did not even send a salesman out when a business showed interest in it; surely, it argued, these people could not possibly know what a computer was all about. IBM was equally convinced that the computer was an instrument for scientific work: their own computer had been designed specifically for astronomical calculations. But IBM was willing to take orders from businesses and to serve them. Ten years later, around 1960, Univac still had by far the most advanced and best machine. IBM had the computer market.

The textbook prescription for this problem is "market research." But it is the wrong prescription.

One cannot do market research for something genuinely new. One cannot do market research for something that is not yet on the market. Similarly, several companies who turned down the Xerox patents did so on the basis of thorough market research, which showed that printers had no use at all for a copier. Nobody had any inkling that businesses, schools, universities, colleges, and a host of private individuals would want to buy a copier.

The new venture therefore needs to start out with the assumption that its product or service may find customers in markets no one thought of, for uses no one envisaged when the product or service was designed, and that it will be bought by customers outside its field of vision and even unknown to the new venture.

To build market focus into a new venture is not in fact particularly difficult. But what is required runs counter to the inclinations of the typical entrepreneur. It requires, first, that the new venture systematically hunt out both the unexpected success and the unexpected failure. Rather than dismiss the unexpected as an "exception," as entrepreneurs are inclined to do, they need to go out and look at it carefully and as a distinct opportunity.

Shortly after World War II, a small Indian engineering firm bought the license to produce a European-designed bicycle with an auxiliary light engine. It looked like an ideal product for India; yet it never did well. The owner of this small firm noticed, however, that substantial orders came in for the engines alone. At first he wanted to turn down those orders; what could anyone possibly do with such a small engine? It was curiosity alone that made him go to the actual area the orders came from. There he found farmers who were taking the engines off the bicycles and using them to power irrigation pumps that hitherto had been hand-operated. This manufacturer is now the world's largest maker of small irrigation pumps, selling them by the millions. His pumps have revolutionized farming all over Southeast Asia.

It does not require a great deal of money to find out whether an unexpected interest from an unexpected market is an indication of genuine potential or a fluke. It requires sensitivity and a little systematic work.

Above all, the people who are running a new venture need to spend time outside: in the marketplace, with customers and with their own salespeople, looking and listening. The new venture needs to build in systematic practices to remind itself that a "product" or a "service" is defined by the customer, not by the producer. It needs to work continually on challenging itself in respect to the utility and value that its products or services contribute to customers.

The greatest danger for the new venture is to "know better" than the customer what the product or service is or should be, how it should be bought, and what it should be used for. Above all, the new venture needs willingness to see the unexpected success as an opportunity rather than as an affront to its expertise. And it needs to accept that elementary axiom of marketing: Businesses are not paid to reform customers. They are paid to satisfy customers.

Financial Foresight

Lack of market focus is typically a disease of the "neonatal," the infant new venture. It is the most serious affliction of the new venture in its early stages—and one that can permanently stunt even those that survive.

The lack of adequate financial focus and of the right financial policies is, by contrast, the greatest threat to the new venture in the next stage of its growth. It is, above all, a threat to the rapidly growing new venture. The more successful a new venture is, the more dangerous the lack of financial foresight.

Suppose that a new venture has successfully launched its product or service and is growing fast. It reports "rapidly increasing profits" and issues rosy forecasts. The stock market then "discovers" the new venture, especially if it is high-tech or in a field otherwise currently fashionable. Predictions abound that the new venture's sales will reach a billion dollars within five years. Eighteen months later the new venture collapses. It may not go out of existence or go bankrupt. But it is suddenly awash in red ink, lays off 180 of its 275 employees, fires the president, or is sold at a bargain price to a big company. The causes are always the same: lack of cash; inability to raise the capital needed for expansion; and loss of control, with expenses, inventories, and receivables in disarray. These three financial afflictions often hit together at the same time. Yet any one of them is enough to endanger the health, if not the life, of the new venture.

Once this financial crisis has erupted, it can be cured only with great difficulty and considerable suffering. But it is eminently preventable.

Entrepreneurs starting new ventures are rarely unmindful of money; on the contrary, they tend to be greedy. They therefore focus on profits. But this is the wrong focus for a new venture, or rather, it should come last rather than first. Cash flow, capital, and controls should be emphasized in the early stages. Without them, the profit figures are fiction—good for twelve to eighteen months, perhaps, after which they evaporate.

Growth has to be fed. In financial terms this means that growth in a new venture demands adding financial resources rather than taking them out. Growth needs more cash and more capital. If the growing new venture shows a "profit," it is a fiction: a bookkeeping entry put in only to balance the accounts. And since taxes are payable on this fiction in most countries, it creates a liability and a cash drain rather than "surplus." The healthier a new venture and the faster it grows, the more financial feeding it requires. The new ventures that are the darlings of the newspapers and the stock market newsletters, the new ventures that show rapid profit growth and "record profits," are those most likely to run into desperate trouble a couple of years later.

The new venture needs cash-flow analysis, cash-flow forecasts, and cash management. The fact that America's new ventures of the last few years (with the significant exception of high-tech companies) have been doing so much better than new ventures used to do is largely because the new entrepreneurs in the United States have learned that entrepreneurship demands financial management.

Cash management is fairly easy if there are reliable cash-flow forecasts, with "reliable," meaning "worst case," assumptions rather than hopes. There is an old banker's rule of thumb, according to which in forecasting cash income and cash outlays one assumes that bills will have to be paid sixty days earlier than expected and receivables will come in sixty days later. If the forecast is overly conservative, the worst that can happen—it rarely does in a growing new venture—is a temporary cash surplus.

A growing new venture should know twelve months ahead of time how much cash it will need, when, and for what purposes. With a year's lead time, it is almost always possible to finance cash needs. But even if a new venture is doing well, raising cash in a hurry and in a "crisis" is never easy and always prohibitively expensive. Above all, it always sidetracks the key people in the company at the most critical time. For several months they then spend their time and energy running from one financial institution to another and cranking out one set of questionable financial projections after another. In the end, they usually have to mortgage the long-range future of the business to get through a ninety-day cash bind. When they finally are able again to devote time and thought to the business, they have irrevocably missed the major opportunities. For the new venture, almost by definition, is under cash pressure when the opportunities are greatest.

The successful new venture will also outgrow its capital structure. A rule of thumb with a good deal of empirical evidence to support it says that a new venture outgrows its capital base with every increase in sales (or billings) of the order of 40 to 50 percent. After such growth, a new venture also needs a new and different capital structure, as a rule. As the venture grows, private sources of funds, whether from the owners and their families or from outsiders, become inadequate. The company has to find access to much larger pools of money by going "public," by finding a partner or partners among established companies, or by raising money from insurance companies and pension funds. A new venture that had been financed by equity money now needs to shift to long-term debt, or vice versa. As the venture grows, the existing capital structure always becomes the wrong structure and an obstacle.

Finally, the new venture needs to plan the financial system it requires to manage growth. Again and again, a growing new venture starts off with an excellent product, excellent standing in its market, and excellent growth prospects. Then suddenly everything goes out of control: receivables, inventory, manufacturing costs, administrative costs, service, distribution, everything. Once one area gets out of control, all of them do. The enterprise has outgrown its control structure. By the time control has been reestablished, markets have been lost, customers have become disgruntled if not hostile, distributors have lost their confidence in the company. Worst of all, employees have lost trust in management, and with good reason.

Fast growth always makes obsolete the existing controls. Again, a growth of 40 to 50 percent in volume seems to be the critical figure.

Once control has been lost, it is hard to recapture. Yet the loss of control can be prevented quite easily. What is needed is first to think through the critical areas in a given enterprise. In one, it may be product quality; in another, service; in a third, receivables and inventory; in a fourth, manufacturing costs. Rarely are there more than four or five critical areas in any given enterprise. (Managerial and administrative overhead should, however, always be included. A disproportionate and fast increase in the percentage of revenues absorbed by managerial and administrative overhead, which means that the enterprise hires managerial and administrative people faster than it actually grows, is usually the first sign that a business is getting out of control, that its management structure and practices are no longer adequate to the task.)

To live up to its growth expectations, a new venture must establish today the controls in these critical areas it will need three years hence. Elaborate controls are not necessary nor does it matter that the figures are only approximate. What matters is that the management of the new venture is aware of these critical areas, is being reminded of them, and can thus act fast if the need arises. Disarray normally does not appear if there is adequate attention to the key areas. Then the new venture will have the controls it needs when it needs them.

Financial foresight does not require a great deal of time. It does require a good deal of thought, however. The technical tools to do the job are easily available; they are spelled out in most texts on managerial accounting. But the work will have to be done by the enterprise itself.

Building a Top Management Team

The new venture has successfully established itself in the right market and has then successfully found the financial structure and the financial system it needs. Nonetheless, a few years later it is still prone to run into a serious crisis. Just when it appears to be on the threshold of becoming an "adult"—a successful, established, going concern—it gets into trouble nobody seems to understand. The products are first-rate, the prospects are excellent, and yet the business simply cannot grow. Neither profitability nor quality, nor any of the other major areas performs.

The reason is always the same: a lack of top management. The business has outgrown being managed by one person, or even two people, and it now needs a management team at the top. If it does not have one already in place at the time, it is very late—in fact, usually too late. The best one can then hope is that the business will survive. But it is likely to be permanently crippled or to suffer wounds that will bleed for many years to come. Morale has been shattered and employees throughout the company are disillusioned and cynical. And the people who founded the business and built it almost always end up on the outside, embittered and disenchanted.

The remedy is simple: to build a top management team *before* the venture reaches the point where it must have one. Teams cannot be formed overnight. They require long periods before they can function. Teams are based on mutual trust and mutual understanding, and this takes years to build up. In my experience, three years is about the minimum.

But the small and growing new venture cannot afford a top management team; it cannot sustain half a dozen people with big titles and corresponding salaries. In fact, in the small and growing business, a very small number of people do everything as it comes along. How, then, can one square this circle?

Again, the remedy is relatively simple. But it does require the will on the part of the founders to build a team rather than to keep on running everything themselves. If one or two people at the top believe that they, and they alone, must do everything, then a management crisis a few months, or at the latest, a few years down the road becomes inevitable.

Whenever the objective economic indicators of a new venture—market surveys, for instance, or demographic analysis—indicate that the business may double within three or five years, then it is the duty of the founder or founders to build the management team the new venture will very soon require. This is preventive medicine, so to speak.

First of all, the founders, together with other key people in the firm, will have to think through the key activities of their business. What are the specific areas upon which the survival and success of this particular business depend? Most of the areas will be on everyone's list. But if there are divergencies and dissents—and there should be on a question as important as this—they should be taken seriously. Every activity that any member of the group thinks belongs there should go down on the list.

The key activities are not to be found in books. They emerge from analysis of the specific enterprise. Two enterprises that to an outsider appear to be in an identical line of business may well end up defining their key activities quite differently. One, for instance, may put production in the center; the other, customer service. Only two key activities are always present in any organization: there is always the management of people and there is always the management of money. The rest has to be determined by the people within looking at the enterprise and at their own jobs, values, and goals.

The next step is, then, for each member of the group, beginning with the founder, to ask: "What are the activities that *I* am doing well? And what are the activities that each of my key associates in this business is actually doing well?" Again, there is going to be agreement on most of the people and on most of their strengths. But, again, any disagreement should be taken seriously.

Next, one asks: "Which of the key activities should each of us, therefore, take on as his or her first and major responsibility because they fit the individual's strengths? Which individual fits which key activity?"

Then the work on building a team can begin. The founder starts to discipline himself (or herself) not to handle people and their problems, if this is not the key activity that fits him best. Perhaps this individual's key strength is new products and new technology. Perhaps this individual's key activity is operations, manufacturing, physical distribution, service. Or perhaps it is money and finance and someone else had better handle people. But all key activities need to be covered by someone who has proven ability in performance.

There is no rule that says, A chief executive has to be in charge of this or that. Of course, a chief executive is the court of last resort and has ultimate accountability. And the chief executive also has to make sure of getting the information necessary to assume this ultimate accountability. The chief executive's own *work*, however, depends on what the enterprise requires and on who the individual is. As long as the CEO's work program consists of key activities, he or she does a CEO's job. But the CEO also is responsible for making sure that all the other key activities are adequately covered.

Finally, goals and objectives for each area need to be set. Everyone who takes on the primary responsibility for a key activity, whether product development or people, or money, must be asked: "What can this enterprise expect of *you?* What should we hold *you* accountable for? What are *you* trying to accomplish and by what time?" But this is elementary management, of course.

It is prudent to establish the top management team informally at first. There is no need to give people titles in a new and growing venture, or to make announcements, or even to pay extra. All this can wait a year or so, until it is clear that the new setup works, and how. In the meantime, all the members of the team have much to learn: their job, how they work together, and what they have to do to enable the CEO and their colleagues to do their jobs. Two or three years later, when the growing venture needs a top management team, it has one.

However, should it fail to provide for a top management structure before it actually needs one, it will lose the capacity to manage itself long before it actually needs to. The founder will have become so overloaded that important tasks will not get done. At that point the company can go one of two ways. The first possibility is that the founder concentrates only on the one or two areas that fit his or her abilities and interests. Although those are key areas, they are not the only crucial ones, and no one will be left to look after the business's other vital areas. Two years later those will have been slighted and the business will be in dire straits. The other, worse, possibility is that the founder is in fact conscientious. He knows that people and money are key areas of concern and need to be taken care of. However, his own abilities and interests, which actually built the business, are in the design and development of new products; but being conscientious, he forces himself to focus on people and finance. Since he is not very gifted in either area, he does poorly in both. It also takes him forever to reach decisions or to do any work in these areas, so that he is forced, by lack of time, to neglect what he is really good at and what the company depends on him for, the development of new technology and new products. Three years later the company will have become an empty shell without the products it needs, but also without the management of people and the management of money it needs.

In the first example, it may be possible to save the company. After all, it has the products. But the founder will inevitably be removed by whoever comes in to salvage the company. In the second case, the company usually cannot be saved at all and has to be sold or liquidated.

Long before it has reached the point where it needs the balance of a top management team, the new venture has to create one. Long before the time has come at which management by one person no longer works and becomes mismanagement, that one person also has to start learning how to work with colleagues, has to learn to trust people, yet also how to hold them accountable. The founder has to learn to become the leader of a team rather than a "star" with "helpers."

"Where Can I Contribute?"

Building a top management team may be the single most important step toward entrepreneurial management in the new venture. It is only the first step, however, for the founders themselves, who then have to think through what their own future is to be.

As a new venture develops and grows, the roles and relationships of the original entrepreneurs inexorably change. If the founders refuse to accept this, they will stunt the business and may even destroy it.

Every founder-entrepreneur nods to this and says, "Amen." Everyone has horror stories of other founder-entrepreneurs who did not change as the venture changed, and who then destroyed both the business and themselves. But even among the founders who can accept that they themselves need to do something, few know how to tackle changing their own roles and relationships. They tend to begin by asking, "What do I like to do?" Or at best, "Where do I fit in?" The right question to start with is, "What will the venture need *objectively* by way of management from here on out?" And in a growing new venture, the founder has to ask that question whenever the business (or the public-service institution) grows significantly or changes direction or character, that is, changes its products, services, markets, or the kind of people it needs.

The next questions the founder must ask are: "What am I good at? What, of all these needs of the venture, could I supply, and supply with distinction?" Only after having thought through those two questions should a founder then ask: "What do I really want to do, and believe in doing? What am I willing to spend years on, if not the rest of my life? Is this something the venture really needs? Is it a major, essential, indispensable contribution?"

But the questions of what a venture needs, what the strengths of the founder-entrepreneur are, and what he or she wants to do might be answered quite differently.

Edwin Land, for instance, the man who invented Polaroid glass and the Polaroid camera, ran the company during the first twelve or fifteen years of its life, until the early 1950s. Then it began to grow fast. Land thereupon designed a top management team and put it in place. As for himself, he decided that he was not the right man for the top management job in the company: what he and he alone could contribute was scientific innovation. Accordingly, Land built himself a laboratory and established himself as the company's consulting director for basic research. The company itself, in its day-to-day operations, was left to others to run.

Ray Kroc, the man who conceived and built McDonald's, reached a similar conclusion. He remained president until he died well past the age of eighty. But he put a top management team in place to run the company and appointed himself the company's "marketing conscience." Until shortly before his death, he visited two or three McDonald's restaurants each week, carefully checking their food quality, the level of cleanliness and friendliness and so on. Above all, he looked at the customers, talked to them, and listened to them. This enabled the company to make the necessary changes to retain its leadership in the fast-food industry.

These questions may not always lead to such happy endings. They may even lead to the decision to leave the company.

In one of the most successful new financial services ventures in the United States, that is what the founder concluded. He did establish a top management team. He asked what the company needed. He looked at himself and his strengths, and he found no match between the needs of the company and his own abilities, let alone between the needs of the company and the things he wanted to do. "I trained my own successor for about eighteen months, then turned the company over to him and resigned," he said. Since then he has started three new businesses, not one of them in finance, has developed them successfully to medium size, and then quit again. He wants to develop new businesses but does not enjoy running them. He accepts that both the businesses and he are better off divorced from one another.

Other entrepreneurs in this same situation might reach different conclusions. The founder of a well-known medical clinic, a leader in its particular field, faced a similar dilemma. The needs of the institution were for an administrator and money-raiser. His own inclinations were to be a researcher and a clinician. But he realized that he was good at raising money and capable of learning to be the chief executive officer of a fairly large health-care organization. "And so," he says, "I felt it my duty to the venture I had created, and to my associates in it, to suppress my own desires and to take on the job of chief administrator and money-raiser. But I would never have done so had I not known that I had the abilities to do the job, and if my advisers and my board had not all assured me that I had these abilities."

The question Where do *I* belong? needs to be faced up to and thought through by the founder-entrepreneur as soon as the venture shows the first signs of success. But the question can be faced up to much earlier. Indeed, it might be best thought through before the new venture is even started.

That is what Soichiro Honda, the founder and builder of Honda Motor Company in Japan, did when he decided to open a small business in the darkest days after Japan's defeat in World War II. He did not start his venture until he had found the right man to be his partner and to run administration, finance, distribution, marketing, sales, and personnel. For Honda had decided from the outset that he belonged in engineering and production and would not run anything else. That decision made the Honda Motor Company.

There is an earlier and even more instructive example, that of Henry Ford. When Ford decided in 1903 to go into business for himself, he did exactly what Honda did forty years later: before starting, he found the right man to be his partner and to run the areas where Ford knew he did not belong—administration, finance, distribution, marketing, sales, and personnel. Like Honda, Henry Ford knew that he belonged in engineering and manufacturing and was going to confine himself to those two areas. The man he found, James Couzens, contributed as much as Ford to the success of the company. Many of the best-known policies and practices of the Ford Motor Company for which Henry Ford is often given credit—the famous five-dollar-a-day wage of 1913, or the pioneering distribution and service policies, for example—were Couzens's ideas and at first resisted by Ford. So effective did Couzens become that Ford grew increasingly jealous of him and forced him out in 1917. The last straw was Couzens's insistence that the Model T was obsolescent and his proposal to use some of the huge profits of the company to start work on a successor.

The Ford Motor Company grew and prospered to the very day of Couzens's resignation. Within a few short months thereafter, as soon as Henry Ford had taken every single top management function into his own hands, forgetting that he had known earlier where he belonged, the Ford Motor Company began its long decline. Henry Ford clung to the Model T for a full ten years, until it had become literally unsalable. And the company's decline was not reversed until thirty years after Couzens's dismissal when, with his grandfather dying, a very young Henry Ford II took over the practically bankrupt business.

The Need for Outside Advice

These last cases point up an important factor for the entrepreneur in the new and growing venture, the need for independent, objective outside advice.

The growing new venture may not need a formal board of directors. Moreover, the typical board of directors very often does not provide the advice and counsel the founder needs. But the founder does need people with whom he can discuss basic decisions and to whom he listens. Such people are rarely to be found within the enterprise. Somebody has to challenge the founder's appraisal of the needs of the venture, and of his own personal strengths. Someone who is not a part of the problem has to ask questions, to review decisions, and above all, to push constantly to have the long-term survival needs of the new venture satisfied by building in the market focus, supplying financial foresight, and creating a functioning top management team. This is the final requirement of entrepreneurial management in the new venture.

The new venture that builds such entrepreneurial management into its policies and practices will become a flourishing large business.

In so many new ventures, especially high-tech ventures, the techniques discussed in this chapter are spurned and even despised. The argument is that they constitute "management" and "We are entrepreneurs." But this is not informality; it is irresponsibility. It confuses manners and substance. It is old wisdom that there is no freedom except under the law. Freedom without law is license, which soon degenerates into anarchy, and shortly thereafter into tyranny. It is precisely because the new venture has to maintain and strengthen the entrepreneurial spirit that it needs foresight and discipline. It needs to prepare itself for the demands its own success will make of it. Above all, it needs responsibility—and this, in the last analysis, is what entrepreneurial management supplies to the new venture.

12.

ENTREPRENEURIAL STRATEGIES

Just as entrepreneurship requires entrepreneurial management, that is, practices and policies within the enterprise, so it requires practices and policies outside, in the marketplace. It requires entrepreneurial strategies.

Of late, "strategy in business" has become the "in" word, with any number of books written about it. However, I have not come across any discussion of entrepreneurial strategies. Yet they are important; they are distinct; and they are different.

There are four specifically entrepreneurial strategies.

1. "Being fustest with the mostest"
2. "Hitting them where they ain't"
3. Finding and occupying a specialized "ecological niche"
4. Changing the economic characteristics of a product, a market, or an industry

These four strategies are not mutually exclusive. One and the same entrepreneur often combines elements of two, sometimes even three, in one strategy. They are also not always sharply differentiated; the same strategy might, for instance, be classified as "hitting them where they ain't" or "finding and occupying a specialized 'ecological niche.'" Still, each of these four has its prerequisites. Each fits certain kinds of innovation and does not fit others. Each requires specific behavior on the part of the entrepreneur. Finally, each has its own limitations and carries its own risks.

"Being Fustest with the Mostest"

"Being fustest with the mostest" was how a Confederate cavalry general in America's Civil War explained consistently winning his battles. In this strategy the entrepreneur aims at leadership, if not at dominance of a new market or a new industry. "Being fustest with the mostest" does not necessarily aim at creating a big business right away, though often this is indeed the aim. But it aims from the start at a permanent leadership position.

"Being fustest with the mostest" is the approach that many people consider the entrepreneurial strategy par excellence. Indeed, if one were to go by the popular books on entrepreneurs, one would conclude that "being fustest with the mostest" is the only entrepreneurial strategy—and a good many entrepreneurs, especially the high-tech ones, seem to be of the same opinion.

They are wrong, however. To be sure, a good many entrepreneurs have indeed chosen this strategy. Yet "being fustest with the mostest" is not even the dominant entrepreneurial strategy, let alone the one with the lowest risk or the highest success ratio. On the contrary, of all entrepreneurial strategies it is the greatest gamble. And it is unforgiving, making no allowances for mistakes and permitting no second chance.

But if successful, "being fustest with the mostest" is highly rewarding.

Here are some examples to show what this strategy consists of and what it requires.

Hoffmann-LaRoche of Basel, Switzerland, has for many years

been the world's largest and in all probability its most profitable pharmaceutical company. But its origins were quite humble: until the mid-1920s, Hoffmann-LaRoche was a small and struggling manufacturing chemist, making a few textile dyes. It was totally overshadowed by the huge German dye-stuff makers and two or three much bigger chemical firms in its own country. Then it gambled on the newly discovered vitamins at a time when the scientific world still could not quite accept that such substances existed. It acquired the vitamin patents—nobody else wanted them. It hired the discoverers away from Zurich University at several times the salaries they could hope to get as professors, salaries even industry had never paid before. And it invested all the money it had and all it could borrow in manufacturing and marketing these new substances.

Sixty years later, long after all vitamin patents have expired, Hoffmann-LaRoche has nearly half the world's vitamin market, now amounting to billions of dollars a year.

Du Pont followed the same strategy. When it came up with nylon, the first truly synthetic fiber, after fifteen years of hard, frustrating research, Du Pont at once mounted massive efforts, built huge plants, went into mass advertising—the company had never before had consumer products to advertise—and created the industry we now call plastics.

Not every "being fustest with the mostest" strategy needs to aim at creating a big business, though it must always aim at creating a business that dominates its market. The 3M Company in St. Paul, Minnesota, does not—as a matter of deliberate policy, it seems—attempt an innovation that might result in a big business by itself. Nor does Johnson & Johnson, the health-care and hygiene producer. Both companies are among the most fertile and most successful innovators. Both look for innovations that will lead to medium-sized rather than to giant enterprises, which are, however, dominant in their markets.

Perhaps because "being fustest with the mostest" must aim at creating something truly new, something truly different, nonexperts and outsiders seem to do as well as the experts, in fact, often better. Hoffmann-LaRoche, for instance, did not owe its strategy to chemists, but to a musician who had married the granddaughter of the company's founder and needed more money to support his orchestra than the company then provided through its meager dividends. To this day the company has never been managed by chemists, but always by financial men who have made their career in a major Swiss bank.

The strategy of "being fustest with the mostest" has to hit right on target or it misses altogether. Or, to vary the metaphor, "being fustest with the mostest" is very much like a moon shot: a deviation of a fraction of a minute of the arc and the missile disappears into outer space. And once launched, the "being fustest with the mostest" strategy is difficult to adjust or to correct.

To use this strategy, in other words, requires thought and careful analysis. The "entrepreneur" who dominates so much of the popular literature or who is portrayed in Hollywood movies, the person who suddenly has a "brilliant idea" and rushes off to put it into effect, is not going to succeed with it.

There has to be one clear-cut goal and all efforts have to be focused on it. And when those efforts begin to produce results, the innovator has to be ready to mobilize resources massively.

Then, after the innovation has become a successful business, the work really begins. Then the strategy of "being fustest with the mostest" demands substantial and continuing efforts to retain a leadership position; otherwise, all one has done is create a market for a competitor. The innovator has to run even harder now that he has leadership than he ran before and to continue his innovative efforts on a very large scale. The research budget must be higher *after* the innovation has successfully been accomplished than it was before. New uses have to be found; new customers must be identified, and persuaded to try the new materials. Above all, the entrepreneur who has succeeded in "being fustest with the mostest" has to make his product or his process obsolete before a competitor can do it. Work on the successor to the successful product or process has to start immediately, with the same concentration of effort and the same investment of resources that led to the initial success.

Finally, the entrepreneur who has attained leadership by "being fustest with the mostest" has to be the one who systematically cuts the price of his own product or process. To keep prices high simply holds an umbrella over potential competitors and encourages them.

The strategy of "being fustest with the mostest" is indeed so risky that an entire major strategy is based on the assumption that "being fustest with the mostest" will fail far more often than it can possibly succeed. It will fail because the will is lacking. It will fail because efforts are inadequate. It will fail because, despite successful innovation, not enough resources are deployed, are available, or are being put to work to exploit success, and so on. While the strategy is indeed highly rewarding when successful, it is much too risky and much too difficult to be used for anything but major innovations.

In most cases alternative strategies are available and preferable— not primarily because they carry less risk, but because for most innovations the opportunity is not great enough to justify the cost, the effort, and the investment of resources required for the "being fustest with the mostest" strategy.

Creative Imitation

Two completely different entrepreneurial strategies were summed up by another battle-winning Confederate general in America's Civil War, who said, "Hit them where they ain't." They might be called creative imitation and entrepreneurial judo, respectively.

Creative imitation is clearly a contradiction in terms. What is creative must surely be original. And if there is one thing imitation is not, it is "original." Yet the term fits. It describes a strategy that is "imitation" in its substance. What the entrepreneur does is something somebody else has already done. But it is "creative" because the entrepreneur applying the strategy of "creative imitation" understands what the innovation represents better than the people who made it and who innovated.

The foremost practitioner of this strategy and the most brilliant one is IBM. And the Japanese Hattori Company, whose Seiko watches have become the world's leader, also owes its domination of the market to creative imitation.

In the early 1930s, IBM built a high-speed calculating machine to do calculations for the astronomers at New York's Columbia University. A few years later it built a machine that was already designed as a computer—again, to do astronomical calculations, this time at Harvard. And by the end of World War II, IBM had built a real computer—the first one, by the way, that had the features of the true computer: a "memory" and the capacity to be "programmed." And yet there are good reasons why the history books pay scant attention to IBM as a computer innovator. For as soon as it had finished its advanced 1945 computer—the first computer to be shown to a lay public in its showroom in midtown New York, where it drew immense crowds—IBM abandoned its own design and switched to the design of its rival, the ENIAC developed at the University of Pennsylvania. The ENIAC was far better suited to business applications such as payroll, only its designers did not see this. IBM structured the ENIAC so that it could be manufactured and serviced and could do mundane "number crunching." When IBM's version of the ENIAC came out in 1953, it at once set the standard for commercial, multipurpose, mainframe computers.

This is the strategy of "creative imitation." It waits until somebody else has established the new, but only "approximately." Then it goes to work. And within a short time it comes out with what the new really should be to satisfy the customer, to do the work customers want and pay for. The creative imitation has then set the standard and takes over the market.

When semiconductors became available, everyone in the watch industry knew that they could be used to power a watch much more accurately, much more reliably, and much more cheaply than traditional watch movements. The Swiss soon brought out a quartz-powered digital watch. But they had so much investment in traditional watchmaking that they decided on a gradual introduction of quartz-powered digital watches over a long period of time, during which these new timepieces would remain expensive luxuries.

Meanwhile, the Hattori Company in Japan had long been making conventional watches for the Japanese market. It saw the opportunity and went in for creative imitation, developing the quartz-powered digital watch as the standard timepiece. By the time the Swiss had woken up, it was too late. Seiko watches had become the world's best sellers, with the Swiss almost pushed out of the market.

Like "being fustest with the mostest," creative imitation is a strategy aimed at market or industry leadership, if not at market or industry dominance. But it is much less risky. By the time the creative imitator moves, the market has been established and the new venture has been accepted. Indeed, there is usually more demand for it than the original innovator can easily supply. The market segmentations are known or at least knowable. By then, too, market research can find out what customers buy, how they buy, what constitutes value for them, and so on.

Of course, the original innovator may do it right the first time, thus closing the door to creative imitation. There is the risk of an innovator bringing out and doing the right job with vitamins as Hoffmann-LaRoche did, or with nylon as did Du Pont. But the number of entrepreneurs engaging in creative imitation, and their substantial success, indicates that perhaps the risk of the first innovator's preempting the market by getting it right is not an overwhelming one.

The creative innovator exploits the success of others. Creative imitation is not "innovation" in the sense in which the term is most commonly understood. The creative imitator does not invent a product or service; he perfects and positions it. In the form in which it has been introduced, it lacks something. It may be additional product features. It may be segmentation of product or services so that slightly different versions fit slightly different markets. It might be proper positioning of the product in the market. Or creative imitation supplies something that is still lacking.

The creative imitator looks at products or services from the viewpoint of the customer. Creative imitation starts out with markets rather than with products, and with customers rather than with producers. It is both market-focused and market-driven. But creative imitators do not succeed by taking away customers from the pioneers who have first introduced a new product or service; they serve markets the pioneers have created but do not adequately service. Creative imitation satisfies a demand that already exists rather than creating one.

The strategy has its own risks, and they are considerable. Creative imitators are easily tempted to splinter their efforts in the attempt to hedge their bets. Another danger is to misread the trend and imitate creatively what then turns out not to be the winning development in the marketplace.

IBM, the world's foremost creative imitator, exemplifies these dangers. It has successfully imitated every major development in the office-automation field. As a result, it has the leading product in every single area. But because they originated in imitation, the products are so diverse and so little compatible with one another that it is all but impossible to build an integrated, automated office out of IBM building blocks. It is thus still doubtful that IBM can maintain leadership in the automated office and provide the integrated system for it. Yet this is where the main market of the future is going to be in all probability. And this risk, *the risk of being too clever,* is inherent in the creative imitation strategy.

Creative imitation is likely to work most effectively in high-tech areas for one simple reason: high-tech innovators are least likely to be market-focused, and most likely to be technology and product-focused. They therefore tend to misunderstand their own success and to fail to exploit and supply the demand they have created.

Entrepreneurial Judo

In 1947, Bell Laboratories invented the transistor. It was at once realized that the transistor was going to replace the vacuum tube, especially in consumer electronics such as the radio and the brand-new television set. Everybody knew this; but nobody did anything about it. The leading manufacturers—at that time they were all Americans—began to study the transistor and to make plans for conversion to the transistor "sometime around 1970." Till then, they proclaimed, the transistor "would not be ready." Sony was practically unknown outside of Japan and was not even in consumer electronics at the time. But Akio Morita, Sony's president, read about the transistor in the newspapers. As a result, he went to the United States and bought a license for the new transistor from Bell Labs for a ridiculous sum, all of twenty-five thousand dollars. Two years later Sony brought out the first portable transistor radio, which weighed less than one-fifth of comparable vacuum tube radios on the market, and cost less than one-third of what they cost. Three years later Sony had the market for cheap radios in the United States; and two years after that, the Japanese had captured the radio market all over the world.

Of course, this is a classic case of the rejection of the unexpected success. The Americans rejected the transistor because it was "not invented here," that is, not invented by one of the electrical and electronic leaders, RCA and GE. It is a typical example of pride in doing things the hard way. The Americans were so proud of the wonderful radios of those days, the great Super Heterodyne sets that were such marvels of craftsmanship. Compared with them, they thought transistors low-grade, if not indeed beneath their dignity.

But Sony's success is not the real story. How do we explain that the Japanese repeated this same strategy again and again, and always with success, always surprising the Americans? The Japanese, in other words, have been successful again and again in practicing "entrepreneurial judo" against the Americans.

But so were MCI and Sprint successful when they used the Bell Telephone System's (AT&T) own pricing to take away from the Bell System a very large part of the long-distance business. So was ROLM when it used Bell System's policies against it to take away a large part of the private branch exchange (PBX) market. And so was Citibank when it started a consumer bank in Germany, the Familienbank (Family Bank), which within a few short years came to dominate German consumer finance.

The German banks knew that ordinary consumers had obtained purchasing power and had become desirable clients. They went through the motions of offering consumers banking services. But they really did not want them. Consumers, they felt, were beneath the dignity of a major bank, with its business customers and its rich investment clients. If consumers needed an account at all, they should have it with the postal savings bank.

All these newcomers—the Japanese, MCI, ROLM, Citibank—practiced "entrepreneurial judo." Of the entrepreneurial strategies, especially the strategies aimed at obtaining leadership and dominance in an industry or a market, entrepreneurial judo is by all odds the least risky and the most likely to succeed.

Every policeman knows that a habitual criminal will always commit his crime the same way—whether it is cracking a safe or entering a building he wants to loot. He leaves behind a "signature," which is as individual and as distinct as a fingerprint. And he will not change that signature even though it leads to his being caught time and again.

But it is not only the criminal who is set in his habits. All of us are. And so are businesses and industries. The habit will be persisted in even though it leads again and again to loss of leadership and loss of market. The American manufacturers persisted in the habits that enabled the Japanese to take over their market again and again.

If the criminal is caught, he rarely accepts that his habit has betrayed him. On the contrary, he will find all kinds of excuses—

and continue the habitual behavior that led to his being captured. Similarly, businesses that are being betrayed by their habits will not admit it and will find all kinds of excuses. The American electronics manufacturers, for instance, attribute the Japanese successes to "low labor costs" in Japan. Yet the few American manufacturers that have faced up to reality, for example, RCA and Magnavox in television sets, are able to turn out in the United States products at prices competitive with those of the Japanese, and competitive also in quality, despite their paying American wages and union benefits. The German banks uniformly explain the success of Citibank's Familienbank by its taking risks they themselves would not touch. But Familienbank has lower credit losses with consumer loans than the German banks, and its lending requirements are as strict as those of the Germans. The German banks know this, of course. Yet they keep on explaining away their failure and Familienbank's success. This is typical. And it explains why the same strategy—the same entrepreneurial judo—can be used over and over again.

There are in particular five fairly common bad habits that enable newcomers to use entrepreneurial judo and to catapult themselves into a leadership position in an industry against the entrenched, established companies.

1. The first is what American slang calls NIH ("not invented here"), the arrogance that leads a company or an industry to believe that something new cannot be any good unless they themselves thought of it. And so the new invention is spurned, as was the transistor by the American electronics manufacturers.

2. The second is the tendency to "cream" a market, that is, to get the high-profit part of it.

This is basically what Xerox did and what made it an easy target for the Japanese imitators of its copying machines. Xerox focused its strategy on the big users, the buyers of large numbers of machines or of expensive, high-performance machines. It did not reject the others; but it did not go after them. In particular, it did not see fit to give them service. In the end it was dissatisfaction with the service—or rather, with the lack of service—Xerox provided for its smaller customers that made them receptive to competitors' machines.

"Creaming" is a violation of elementary managerial and economic precepts. It is always punished by loss of market.

3. Even more debilitating is the third bad habit: the belief in "quality." "Quality" in a product or service is not what the supplier puts in. It is what the customer gets out and is willing to pay for. A product is not "quality" because it is hard to make and costs a lot of money, as manufacturers typically believe. That is incompetence. Customers pay only for what is of use to them and gives them value. Nothing else constitutes "quality."

4. Closely related to both "creaming" and "quality" is the fourth bad habit, the illusion of the "premium" price. A "premium" price is always an invitation to the competitor.

What looks like higher profits for the established leader is in effect a subsidy to the newcomer who, in a very few years, will unseat the leader and claim the throne for himself. "Premium" prices, instead of being an occasion for joy—and a reason for a higher stock price or a higher price/earnings multiple—should always be considered a threat and a dangerous vulnerability.

Yet the illusion of higher profits to be achieved through "premium" prices is almost universal, even though it always opens the door to entrepreneurial judo.

5. Finally, there is a fifth bad habit that is typical of established businesses and leads to their downfall. They maximize rather than optimize. As the market grows and develops, they try to satisfy every single user through the same product or service. Xerox is a good example of a company with this habit.

Similarly, when the Japanese came onto the market with their copiers in competition with Xerox, they designed machines that fitted specific groups of users—for example, the small office, whether that of the dentist, the doctor, or the school principal. They did not try to match the features of which the Xerox people were the proudest, such as the speed of the machine or the clarity of the copy. They gave the small office what the small office needed most, a simple machine at a low cost. And once they had established themselves in that market, they then moved in on the other markets, each with a product designed to serve optimally a specific market segment.

Sony similarly first moved into the low end of the radio market, the market for cheap portables with limited range. Once it had established itself there, it moved in on the other market segments.

Entrepreneurial judo aims first at securing a beachhead, and one that the established leaders either do not defend at all or defend only halfheartedly—the way the Germans did not counterattack when Citibank established its Familienbank. Once that beachhead has been secured, that is, once the newcomers have an adequate market and an adequate revenue stream, they then move on to the rest of the "beach" and finally to the whole "island." In each case, they repeat the strategy. They design a product or a service that is specific to a given market segment and optimal for it. And the established leaders hardly ever beat them at this game. Hardly ever do the established leaders manage to change their own behavior before the newcomers have taken over the leadership and acquired dominance.

Entrepreneurial judo requires some degree of genuine innovation. It is, as a rule, not good enough to offer the same product or the same service at lower cost. There has to be something that distinguishes it from what already exists.

It is not enough, in other words, for the newcomer simply to do as good a job as the established leader at a lower cost or with better service. The newcomers have to make themselves distinct.

Like "being fustest with the mostest" and creative imitation, entrepreneurial judo aims at obtaining leadership position and eventually dominance. But it does not do so by competing with the leaders—or at least not where the leaders are aware of competitive challenge or worried about it. Entrepreneurial judo "hits them where they ain't."

The Toll-Gate Strategy

The entrepreneurial strategies discussed so far, "being fastest with the mostest," "hitting them where they ain't" or creative imitation, and entrepreneurial judo, all aim at market or industry leadership, if not at dominance. The "ecological niche" strategy aims at control. The strategies discussed earlier aim at positioning an enterprise in a large market or a major industry. The ecological niche strategy aims at obtaining a practical monopoly in a small area. The first three strategies are competitive strategies. The ecological niche strategy aims at making its successful practitioners immune to competition and unlikely to be challenged. Successful practitioners of "being fastest with the mostest," creative imitation, and entrepreneurial judo become big companies, highly visible if not household words. Successful practitioners of the ecological niche take the cash and let the credit go. They revel in their anonymity. Indeed, in the most successful uses of the ecological niche strategy, the whole point is to be so inconspicuous, despite the product's being essential to a process, that no one is likely to try to compete.

There are three distinct versions of the ecological niche strategy, each with its own requirements, its own limitations, and its own risks.

- ➤ Toll-gate strategy
- ➤ Specialty skill strategy
- ➤ Specialty market strategy

The Alcon Company developed an enzyme to eliminate the one feature of the standard surgical operation for senile cataracts that went counter to the rhythm and the logic of the process. Once this enzyme had been developed and patented, it had a "toll-gate" position. No eye surgeon would do without it. No matter what Alcon charged for the teaspoonful of enzyme that was needed for each cataract operation, the cost was insignificant in relation to the total cost of the operation. I doubt that any eye surgeon or any hospital ever even inquired what the stuff cost. The total market for this particular preparation was so small—maybe fifty million dollars a year worldwide—that it clearly would not have been worth anybody's while to try to develop a competing product. There would not have been one additional cataract operation in the world just because this particular enzyme had become cheaper. All that potential competitors could possibly do, therefore, would have been to knock down the price for everybody, without deriving much benefit for themselves.

The toll-gate position is thus in many ways the most desirable position a company can occupy. But it has stringent requirements. The product has to be essential to a process. The risk of not using it—the risk of losing an eye—must be infinitely greater than the cost of the product. The market must be so limited that whoever occupies it first preempts it. It must be a true "ecological niche" that one species fills completely, and which at the same time is small and discreet enough not to attract rivals.

Such toll-gate positions are not easily found. Normally they occur only in an incongruity situation. The incongruity, as in the case of Alcon's enzyme, might be an incongruity in the rhythm or the logic of a process.

The toll-gate position also has severe limitations and serious risks. It is basically a static position. Once the ecological niche has been occupied, there is unlikely to be much growth. There is nothing the company that occupies the toll-gate position can do to increase its business or to control it. No matter how good its product or how cheap, the demand is dependent upon the demand for the process or product to which the toll-gate product furnishes an ingredient.

Once the toll-gate strategy has attained its objective, the company is "mature." It can only grow as fast as its end users grow. But it can go down fast. It can become obsolete almost overnight if someone finds a different way of satisfying the same end use.

And the toll-gate strategist must never exploit his monopoly. He must not become what the Germans call a *Raubritter* (the English "robber baron" does not mean quite the same thing), who robbed and raped the hapless travelers as they passed through the mountain defiles and river gorges atop of which perched his castle. He must not abuse his monopoly to exploit, extort, or maltreat his customers. If he does, the users will put another supplier into business, or they will switch to less effective substitutes that they can then control.

The Specialty Skill Strategy

Everybody knows the major automobile nameplates. But few people know the names of the companies that supply the electrical and lighting systems for these cars, and yet there are far fewer such systems than there are automobile nameplates: in the United States, the Delco group of GM; in Germany, Robert Bosch; in Great Britain, Lucas; and so on.

But once these companies had attained their controlling position in their specialty skill niche, they retained it. Unlike the toll-gate companies, theirs is a fairly large niche, yet it is still unique. It was obtained by developing high skill at a very early time. An enterprising German attained such a hold on one specialty skill niche that guidebooks for tourists are still called by his name, "Baedeker."

As these cases show, timing is of the essence in establishing a specialty skill niche. It has to be done at the very beginning of a new industry, a new custom, a new market, a new trend. Karl Baedeker published his first guidebook in 1828, as soon as the first steamships on the Rhine opened tourist travel to the middle classes. He then had the field virtually to himself until World War I made German books unacceptable in Western countries.

To attain a specialty niche always requires something new, something added, something that is genuine innovation. There were guidebooks for travelers before Baedeker, but they confined themselves to the cultural scene—churches, sights, and so on. For practical details—the hotels, the fare of the horse-drawn cabs, the distances, and the proper amount to tip—the traveling English milord relied on a professional, the courier. But the middle class had no courier, and that was Baedeker's opportunity. Once he had learned what information the traveler needed, how to get at it and to present it (the format he established is still the one many guidebooks follow), it would not have paid anyone to duplicate Baedeker's investment and build a competing organization.

In the early stages of a major new development, the specialty skill niche offers an exceptional opportunity. Examples abound. For many, many years there were only two companies in the United States making airplane propellers, for instance. Both had been started before World War I.

A specialty skill niche is rarely found by accident. In every case, it results from a systematic survey of innovative opportunities. In every case, the entrepreneur looks for the place where a specialty skill can be developed and can give a new enterprise a unique controlling position.

Robert Bosch spent years studying the new automotive field to position his new company where it could immediately establish itself as the leader. Hamilton Propeller, for many years the leading airplane propeller manufacturer in the United States, was the result of a systematic search by its founder in the early days of powered flight. Baedeker made several attempts to start a service for the tourist before he decided on the guidebook that then bore his name and made him famous.

The first point, therefore, is that in the early stages of a new industry, a new market, or a new major trend, there is the opportunity to search systematically for the specialty skill opportunity—and then there is usually time to develop a unique skill.

The second point is that the specialty skill niche does require a skill that is both unique and different. The early automobile pioneers were, without exception, mechanics. They knew a great deal about machinery, about metals, and about engines. But electricity was alien to them. It required theoretical knowledge that they neither possessed nor knew how to acquire. There were other publishers in Baedeker's time, but a guidebook that required on-the-spot gathering of an enormous amount of detailed information, constant inspection, and a staff of traveling auditors was not within their purview.

The business that establishes itself in a specialty skill niche is therefore unlikely to be threatened by its customers or by its suppliers. Neither of them really wants to get into something that is so alien in skill and in temperament.

Third, a business occupying a specialty skill niche must constantly work on improving its own skill. It has to stay ahead. Indeed, it has to make itself constantly obsolete. The automobile companies in the early days used to complain that Delco in Dayton, and Bosch in Stuttgart, were pushing them. They turned out lighting systems that were far ahead of the ordinary automobile, ahead of what the automobile manufacturers of the times thought the customer needed, wanted, or could pay for, ahead very often of what the automobile manufacturer knew how to assemble.

While the specialty skill niche has unique advantages, it also has severe limitations. One is that it inflicts tunnel vision on its occupants. In order to maintain themselves in their controlling position, they have to learn to look neither right nor left, but directly ahead at their narrow area, their specialized field.

A second, serious limitation is that the occupant of a specialty skill niche is usually dependent on somebody else to bring his product or service to market. It becomes a component. The strength of the automobile electrical firms is that the customer does not know that they exist. But this is of course also their weakness.

Finally, the greatest danger to the specialty skill niche manufacturer is for the specialty to cease being a specialty and to become universal.

The specialty skill niche, like all ecological niches, is limited—in scope as well as in time. Species that occupy such a niche, biology teaches, do not easily adapt to even small changes in the external environment. And this is true, too, of the entrepreneurial skill species. But within these limitations, the specialty skill niche is a highly advantageous position. In a rapidly expanding new technology, industry, or market, it is perhaps the most advantageous strategy. Very few of the automobile makers of 1920 are still around; every one of the electrical and lighting systems makers is. Once attained and properly maintained, the specialty skill niche protects against competition, precisely because no automobile buyer knows or cares who makes the headlights or the brakes. No automobile buyer is therefore likely to shop around for either. Once the name "Baedeker" had become synonymous with tourist guidebooks, there was little danger that anybody else would try to muscle in, at least not until the market changed drastically. In a new technology, a new industry, or a new market, the specialty skill strategy offers an optimal ratio of opportunity to risk of failure.

The Specialty Market Strategy

The major difference between the specialty skill niche and the specialty market niche is that the former is built around a product or service and the latter around specialized knowledge of a market. Otherwise, they are similar.

Two medium-sized companies, one in northern England and one in Denmark, supply the great majority of the automated baking ovens for cookies and crackers bought in the non-Communist world. For many decades, two companies—the two earliest travel agents, Thomas Cook in Europe and American Express in the United States—had a practical monopoly on traveler's checks.

There is, I am told, nothing very difficult or particularly technical about baking ovens. There are literally dozens of companies around that could make them just as well as those two firms in England and Denmark. But these two know the market: they know every major baker, and every major baker knows them. The market is just not big enough or attractive enough to try to compete with these two, as long as they remain satisfactory. Similarly, traveler's checks were a backwater business until the post–World War II period of mass travel. They were highly profitable since the issuer, whether Cook or American Express, had the use of the money and kept the interest earned on it until the purchaser cashed the check—sometimes months after the checks were purchased. But the market was not large enough to tempt anyone else. Furthermore, traveler's checks required a worldwide organization, which Cook and American Express had to maintain anyhow to service their travel customers, and which nobody else in those days had any reason to build.

The specialty market is found by looking at a new development with the question, What opportunities are there in this that would give us a unique niche, and what do we have to do to fill it ahead of everybody else? The traveler's check was no great "invention." It was basically nothing more than a letter of credit, and that had been around for hundreds of years. What was new was that traveler's checks were offered—at first to the customers of Cook and American Express, and then to the general public—in standard denominations. And they could be cashed wherever Cook or American Express had an office or an agent. That made them uniquely attractive to the tourist who did not want to carry a great deal of cash and did not have the established banking connections to make them eligible for a letter of credit.

There was nothing particularly advanced in the early baking ovens, nor is there any high technology in the baking ovens installed today. What the two leading firms did was to realize that the act of baking cookies and crackers was moving out of the home and into the factory. They then studied what commercial bakers needed so that they could manufacture the product their own customers, grocers and supermarkets, could in turn sell and the housewife would buy. The baking ovens were not based on engineering but on market research; the engineering would have been available to anyone.

The specialty market niche has the same requirements as the

specialty skill niche: systematic analysis of a new trend, industry, or market; a specific innovative contribution, if only a "twist" like the one that converted the traditional letter of credit into the modern traveler's check; and continual work to improve the product and especially the service, so that leadership, once obtained, will be retained.

And it has the same limitations. The greatest threat to the specialty market position is success. The greatest threat is when the specialty market becomes a mass market.

Traveler's checks have now become a commodity and highly competitive because travel has become a mass market.

So have perfumes. A French firm, Coty, created the modern perfume industry. It realized that World War I had changed the attitude toward cosmetics. Whereas before the war only "fast women" used cosmetics—or dared admit to their use—cosmetics had become accepted and respectable. By the mid-1920s, Coty had established itself in what was almost a monopoly position on both sides of the Atlantic. Until 1929 the cosmetics market was a "specialty market," a market of the upper middle class. But then during the Great Depression it exploded into a genuine mass market. It also split into two segments: a prestige segment, with high prices, specialty packaging, and specialty distribution; and popular-priced, mass brands sold in every outlet including the supermarket, the variety store, and the drugstore. Within a few short years, the specialty market dominated by Coty had disappeared. But Coty could not make up its mind whether to try to become one of the mass marketers in cosmetics or one of the luxury producers. It tried to stay in a market that no longer existed, and has been drifting ever since.

Creating Customer Utility

In the entrepreneurial strategies discussed so far in this chapter, the aim is to introduce an innovation. In the entrepreneurial strategy discussed in this section, the strategy itself is the innovation. The product or service it carries may well have been around a long time. But the strategy converts this old, established product or service into something new. It changes its utility, its value, its economic characteristics. While physically there is no change, economically there is something different and new.

All the strategies discussed in this section have one thing in common. They create a customer—and that is the ultimate purpose of a business, indeed, of economic activity. But they do so in four different ways:

> Creating utility
> Pricing
> Adaptation to the customer's social and economic reality
> Delivering what represents true value to the customer

Price is usually almost irrelevant in the strategy of creating utility. The strategy works by enabling customers to do what serves *their purpose*. It works because it asks, What is truly a "service," truly a "utility" to the customer?

Every American bride wants to get one set of "good china." A whole set is, however, far too expensive a present, and the people giving her a wedding present do not know what pattern the bride wants or what pieces she already has. So they end up giving something else. The demand was there, in other words, but the utility was lacking. A medium-sized dinnerware manufacturer, the Lenox China Company, saw this as an innovative opportunity. Lenox adapted an old idea, the "bridal register," so that it only "registers" Lenox china. The bride-to-be then picks one merchant whom she tells what pattern of Lenox china she wants, and to whom she refers potential donors of wedding gifts. The merchant then asks the donor, "How much do you want to spend?" and explains, "That will get you two coffee cups with saucers." Or the merchant can say, "She already has all the coffee cups; what she needs now is dessert plates." The result is a happy bride, a happy wedding-gift donor, and a very happy Lenox China Company.

Again, there is no high technology here, nothing patentable, nothing but a focus on the needs of the customer. Yet the bridal register, for all its simplicity—or perhaps because of it—has made Lenox the favorite "good china" manufacturer and one of the most rapidly growing of medium-sized American manufacturing companies.

Pricing

For many years the best-known American face in the world was that of King Gillette, which graced the wrapper of every Gillette razor blade sold anyplace in the world. And millions of men all over the world used a Gillette razor blade every morning.

King Gillette did not invent the safety razor; dozens of them were patented in the closing decades of the nineteenth century.

Gillette's safety razor was no better than many others, and it was a good deal more expensive to produce. But Gillette did not "sell" the razor. He practically gave it away by pricing it at fifty-five cents retail or twenty cents wholesale, not much more than one-fifth of its manufacturing cost. But he designed it so that it could use only his patented blades. These cost him less than one cent apiece to make: he sold them for five cents. And since the blades could be used six or seven times, they delivered a shave at less than one cent apiece—or at less than one-tenth the cost of a visit to a barber.

What Gillette did was to price what the customer buys, namely, the shave, rather than what the manufacturer sells. In the end, the captive Gillette customer may have paid more than he would have paid had he bought a competitor's safety razor for five dollars, and then bought the competitor's blades selling at one cent or two. Gillette's customers surely knew this; customers are more intelligent than either advertising agencies or Ralph Nader believe. But Gillette's pricing made sense to them. They were paying for what they bought, that is, for a shave, rather than for a "thing." And the shave they got from the Gillette razor and the Gillette razor blade was much more pleasant than any shave they could have given themselves with that dangerous weapon, the straight-edge razor, and far cheaper than they could have gotten at the neighborhood barber's.

One reason why the patents on a copying machine ended up at a small, obscure company in Rochester, New York, then known as the Haloid Company, rather than at one of the big printing-machine manufacturers, was that none of the large established manufacturers saw any possibility of selling a copying machine. Their calculations showed that such a machine would have to sell for at least $4,000. Nobody was going to pay such a sum for a copying machine when carbon paper cost practically nothing. Also, of course, to spend $4,000 on a machine meant a capital-appropriations request, which had to go all the way up to the board of directors accompanied by a calculation showing the return on investment, both of which seemed unimaginable for a gadget to help the secretary. The Haloid Company—the present Xerox—did a good deal of technical work to design the final machine. But its major contribution was in pricing. It did not sell the machine; it sold what the machine produced, copies. At five or ten cents a copy, there is no need for a capital-appropriations request. This is "petty cash," which the secretary can disburse without going upstairs. Pricing the Xerox machine at five cents a copy was the true innovation.

Most suppliers, including public-service institutions, never think of pricing as a strategy. Yet pricing enables the customer to pay for what he buys—a shave, a copy of a document—rather than for what the supplier makes. What is being paid in the end is, of course, the same amount. But how it is being paid is structured to the needs and the realities of the consumer. It is structured in accordance with what the consumer actually buys. And it charges for what represents "value" to the customer rather than what represents "cost" to the supplier.

The Customer's Reality

The worldwide leadership of the American General Electric Company (GE) in large steam turbines is based on GE's having thought through, in the years before World War I, what its customers' realities were. Steam turbines, unlike the piston-driven steam engines, which they replaced in the generation of electric power, are complex, requiring a high degree of engineering in their design, and skill in building and fitting them. This the individual electric power company simply cannot supply. It buys a major steam turbine maybe every five or ten years when it builds a new power station. Yet the skill has to be kept in readiness all the time. The manufacturer, therefore, has to set up and maintain a massive consulting organization.

But, as GE soon found out, the customer cannot pay for consulting services. Under American law, the state public utility commissions would have to allow such an expenditure. In the opinion of the commissions, however, the companies should have been able to do this work themselves. GE also found that it could not add to the price of the steam turbine the cost of the consulting services that its customers needed. Again, the public utility commissions would not have accepted it. But while a steam turbine has a very long life, it needs a new set of blades fairly often, maybe every five to seven years, and these blades have to come from the maker of the original turbine. GE built up the world's foremost consulting engineering organization on electric power stations—though it was careful not to call this consulting engineering but "apparatus sales"—for which it did not charge. Its steam turbines were no more expensive than those of its competitors. But it put the added cost of the consulting organization plus a substantial profit into the price it charged for replacement blades. Within ten years all the other manufacturers of steam turbines had caught on and switched to the same system. But by then GE had world market leadership.

Much earlier, during the 1840s, a similar design of product and process to fit customer realities led to the invention of installment buying. Cyrus McCormick was one of many Americans who built a harvesting machine—the need was obvious. And he found, as had the other inventors of similar machines, that he could not sell his product. The farmer did not have the purchasing power. That the machine would earn back what it cost within two or three seasons, everybody knew and accepted, but there was no banker then who would have lent the American farmer the money to buy a machine. McCormick offered installments, to be paid out of the savings the harvester produced over the ensuing three years. The farmer could now afford to buy the machine—and he did so.

Manufacturers are wont to talk of the "irrational customer" (as do economists, psychologists, and moralists). But there are no "irrational customers." As an old saying has it, There are only lazy manufacturers. The customer has to be assumed to be rational. His or her reality, however, is usually quite different from that of the manufacturer.

Delivering Value to the Customer

The last of these innovative strategies delivers what is "value" to the customer rather than what is "product" to the manufacturer. It is actually only one step beyond the strategy of accepting the customer's reality as part of the product and part of what the customer buys and pays for.

A medium-sized company in America's Midwest supplies more than half of all the special lubricant needed for very large earth-moving and hauling machines: the bulldozers and draglines used by contractors building highways; the heavy equipment used to remove the overlay from strip mines; the heavy trucks used to haul coal out of coal mines; and so on. This company is in competition with some of the largest oil companies, which can mobilize whole battalions of lubrication specialists. It competes by not selling lubricating oil at all. Instead, it sells what is, in effect, insurance. What is "value" to the contractor is not lubrication: it is operating the equipment. Every hour the contractor loses because this or that piece of heavy equipment cannot operate costs him infinitely more than he spends on lubricants during an entire year. In all its activities there is a heavy penalty for contractors who miss their deadlines—and they can only get the contract by calculating the deadline as finely as possible and racing against the clock. What the midwestern lubricant maker does is to offer contractors an analysis of the maintenance needs of their equipment. Then it offers them a maintenance program with an annual subscription price, and guarantees the subscribers that their heavy equipment will not be shut down for more than a given number of hours per year because of lubrication problems. Needless to say, the program always prescribes the manufacturer's lubricant. But that is not what contractors buy. They are buying trouble-free operations, which are extremely valuable to them.

These examples are likely to be considered obvious. Surely, anybody applying a little intelligence would have come up with these and similar strategies? But the father of systematic economics, David Ricardo, is believed to have said once, "Profits are not made by differential cleverness, but by differential stupidity." The strategies work, not because they are clever, but because most suppliers—of goods as well as of services, businesses as well as public-service institutions—do not think. They work precisely because they are so "obvious." Why, then, are they so rare? For, as these examples show, anyone who asks the question, What does the customer really buy? will win the race. In fact, it is not even a race since nobody else is running. What explains this?

One reason is the economists and their concept of "value." Every economics book points out that customers do not buy a "product," but what the product does for them. And then, every economics book promptly drops consideration of everything except the "price" for the product, a "price" defined as what the customer pays to take possession or ownership of a thing or a service. What the product does for the customer is never mentioned again. Unfortunately, suppliers, whether of products or of services, tend to follow the economists.

It is meaningful to say that "product A costs X dollars." It is meaningful to say that "we have to get Y dollars for the product to cover our own costs of production and have enough left over to cover the cost of capital, and thereby to show an adequate profit." But it makes no sense at all to conclude, ". . . and therefore the customer has to pay the lump sum of Y dollars in cash for each piece of product A he buys." Rather, the argument should go as follows: "What the customer pays for each piece of the product has to work out as Y dollars *for us*. But how the customer pays depends on what makes the most sense to him. It depends on what the product does for the customer. It depends on what fits his reality. It depends on what the customer sees as 'value.' "

Price in itself is not "pricing," and it is not "value."

But this is nothing but elementary marketing, most readers will protest, and they are right. It is *nothing* but elementary marketing. To start out with the customer's utility, with what the customer buys, with what the realities of the customer are and what the customer's values are—this is what marketing is all about. But why, after forty years of preaching marketing, teaching marketing, professing marketing, so few suppliers are willing to follow, I cannot explain. The fact remains that so far, anyone who is willing to use marketing as the basis for strategy is likely to acquire leadership in an industry or a market fast and almost without risk.

II.

THE INDIVIDUAL

13.

EFFECTIVENESS MUST BE LEARNED

To be effective is the job of the knowledge worker. Whether he or she works in a business or in a hospital, in a government agency or in a labor union, in a university or in the army, the knowledge worker is, first of all, expected to *get the right things done.* And this means simply that the knowledge worker is expected to be effective.

Yet people of high effectiveness are conspicuous by their absence in knowledge jobs. High intelligence is common enough among knowledge workers. Imagination is far from rare. The level of knowledge tends to be high. But there seems to be little correlation between a man's effectiveness and his intelligence, his imagination, or his knowledge. Brilliant men are often strikingly ineffectual; they fail to realize that the brilliant insight is not by itself achievement. They never have learned that insights become effectiveness only through hard systematic work. Conversely, in every organization there are some highly effective plodders. While others rush around in the frenzy and busyness that very bright people so often confuse with "creativity," the plodder puts one foot in front of the other and gets there first, like the tortoise in the old fable.

Intelligence, imagination, and knowledge are essential resources, but only effectiveness converts them into results. By themselves, they only set limits to what can be attained.

Why We Need Effectiveness

All this should be obvious. But why then has so little attention been paid to effectiveness, in an age in which there are mountains of books and articles on every other aspect of the executive's tasks?

One reason for this neglect is that effectiveness is the specific technology of the knowledge worker within an organization. Until recently, there was no more than a handful of these around.

For manual work, we need only efficiency, that is, the ability to do things right rather than the ability to get the right things done. The manual worker can always be judged in terms of the quantity and quality of a definable and discrete output, such as a pair of shoes. We have learned how to measure efficiency and how to define quality in manual work during the last hundred years—to the point where we have been able to multiply the output of the individual worker tremendously.

Formerly, the manual worker—whether machine operator or front-line soldier—predominated in all organizations. Few people of effectiveness were needed: mainly those at the top who gave the orders that others carried out. They were so small a fraction of the total work population that we could, rightly or wrongly, take their effectiveness for granted. We could depend on the supply of "naturals," the few people in any area of human endeavor who somehow know what the rest of us have to learn the hard way.

In fact, only a small fraction of the knowledge workers of earlier days were part of an organization. Most of them worked by themselves as professionals, at most with an assistant. Their effectiveness or lack of effectiveness concerned only themselves and affected only themselves.

Today, however, the large knowledge organization is the central reality. Modern society is a society of large organized institutions. In every one of them, including the armed services, the center of gravity has shifted to the knowledge worker, the man who puts to work what he has between his ears rather than the brawn of his muscles or the skill of his hands. Increasingly, the majority of people who have been schooled to use knowledge, theory, and concept rather than physical force or manual skill work in an organization and are effective insofar as they can make a contribution to the organization.

Now effectiveness can no longer be taken for granted. Now it can no longer be neglected.

The imposing system of measurements and tests that we have developed for manual work—from industrial engineering to quality control—is not applicable to knowledge work. There are few things less pleasing to the Lord, and less productive, than an engineering department that rapidly turns out beautiful blueprints for the wrong product. Working on the *right* things is what makes knowledge work effective. This is not capable of being measured by any of the yardsticks for manual work.

Knowledge workers cannot be supervised closely or in detail. They can only be helped. But they must direct themselves, and they must do so toward performance and contribution, that is, toward effectiveness.

A cartoon in *The New Yorker* magazine showed an office on the door of which was the legend: CHAS. SMITH, GENERAL SALES MANAGER, AJAX SOAP COMPANY. The walls were bare except for a big sign saying, THINK. The man in the office had his feet propped up on his desk and was blowing smoke rings at the ceiling. Outside two older men went by, one saying to the other: "But how can we be sure that Smith thinks soap?"

One can indeed never be sure what the knowledge worker thinks—and yet thinking is his or her specific work; it is the knowledge worker's "doing."

The motivation of the knowledge worker depends on his being effective, on being able to achieve. If effectiveness is lacking in his work, his commitment to work and to contribution will soon wither, and he will become a time-server going through the motions from nine to five.

The knowledge worker does not produce something that is effective by itself. He does not produce a physical product—a ditch, a pair of shoes, a machine part. He produces knowledge, ideas, information. By themselves these "products" are useless. Somebody else, another person of knowledge, has to take them as his input and convert them into his output before they have any reality. The greatest wisdom not applied to action and behavior is meaningless data. The knowledge worker, therefore, must do something that a manual worker need not do. He must provide effectiveness. He cannot depend on the utility his output carries with it as does a well-made pair of shoes.

The knowledge worker is the one "factor of production" through which the highly developed societies and economies of today—the United States, Western Europe, Japan, and also increasingly, the Soviet Union—become and remain competitive.

Who Is an Executive?

Every knowledge worker in a modern organization is an "executive" if, by virtue of his position or knowledge, he or she is responsible for a contribution that materially affects the capacity of the organization to perform and to obtain results. This may be the capacity of a business to bring out a new product or to obtain a larger share of a given market. It may be the capacity of a hospital to provide bedside care to its patients, and so on. Such a man or woman must make decisions; he cannot just carry out orders. He must take responsibility for his contribution. And he is supposed, by virtue of his knowledge, to be better equipped to make the right decision than anyone else. He may be overridden; he may be demoted or fired. But so long as he has the job, the goals, the standards, and the contribution are in his keeping.

This fact is perhaps best illustrated by a recent newspaper interview with a young American infantry captain in the Vietnam jungle.

Asked by the reporter, "How in this confused situation can you retain command?" the young captain said, "Around here, I am only the guy who is responsible. If these men don't know what to do when they run into an enemy in the jungle, I'm too far away to tell them. My job is to make sure they know. What they do depends on the situation which only they can judge. The responsibility is always mine, but the decision lies with whoever is on the spot."

In a guerrilla war, every person is an "executive."

Knowledge work is not defined by quantity. Neither is knowledge work defined by its costs. Knowledge work is defined by its results. And for these, the size of the group and the magnitude of the managerial job are not even symptoms.

Having many people working in market research may endow the results with that increment of insight, imagination, and quality that gives a company the potential of rapid growth and success. If so, two hundred people are cheap. But it is just as likely that the manager will be overwhelmed by all the problems two hundred people bring to their work and cause through their interactions. He may be so busy "managing" as to have no time for market research and for fundamental decisions. He may be so busy checking figures that he never asks the question, What do we really mean when we say "our market"? And as a result, he may fail to notice significant changes in the market that eventually may cause the downfall of his company.

But the individual market researcher without a staff may be equally productive or unproductive. He may be the source of the knowledge and vision that make his company prosper. Or he may spend so much of his time hunting down details—the footnotes academicians so often mistake for research—as to see and hear nothing and to think even less.

Throughout every one of our knowledge organizations, we have people who manage no one and yet are executives. Rarely indeed do we find a situation such as that in the Vietnam jungle, where at any moment, any member of the entire group may be called upon to make decisions with life-and-death impact for the whole. But the chemist in the research laboratory who decides to follow one line of inquiry rather than another one may make the entrepreneurial decision that determines the future of his company. He may be the research director. But he also may be—and often is—a chemist with no managerial responsibilities, if not even a fairly junior employee. Similarly, the decision what to consider one "product" in the account books may be made by a senior vice president in the company. It may also be made by a junior. And this holds true in all areas of today's large organization.

I have called "executives" those knowledge workers, managers, or individual professionals who are expected by virtue of their position or their knowledge to make decisions in the normal course of their work that have impact on the performance and results of the whole. What few yet realize, however, is how many people there are even in the most humdrum organization of today, whether business or government agency, research lab or hospital, who have to make decisions. For the authority of knowledge is surely as legitimate as the authority of position. These decisions, moreover, are of the same *kind* as the decisions of top management.

The most subordinate, we now know, may do the same kind of work as the president of the company or the administrator of the government agency, that is, plan, organize, integrate, motivate, and measure. His compass may be quite limited, but within his sphere, he is an executive.

Similarly, every decision-maker does the same kind of work as the company president or the administrator. His scope may be quite limited. But he is an executive even if his function or his name appears neither on the organization chart nor in the internal telephone directory.

And whether chief executive or beginner, he needs to be effective.

Executive Realities

The realities of the knowledge workers' situation both demand effectiveness from them and make effectiveness exceedingly difficult to achieve. Indeed, unless they work at becoming effective, the realities of their situation will push them into futility.

In their situation there are four major realities over which they essentially no control. Every one of them is built into the organization and into the executives' day and work. They have no choice but to "cooperate with the inevitable." But every one of these realities exerts pressure toward nonresults and nonperformance.

1. The executive's time tends to belong to everybody else. If one attempted to define an "executive" operationally (that is, through his activities), one would have to define him as a captive of the organization. Everybody can move in on his time, and everybody does. There seems to be very little any one executive can do about it. He cannot, as a rule, like the physician, stick his head out the door and say to the nurse, "I won't see anybody for the next half hour." Just at this moment, the executive's telephone rings, and he has to speak to the company's best customer or to a high official in the city administration or to his boss—and the next half hour is already gone.

2. Executives are forced to keep on "operating" unless they take positive action to change the reality in which they live and work.

But events rarely tell the executive anything, let alone the real problem. For the doctor, the patient's complaint is central because it is central to the patient. The executive is concerned with a much more complex universe. What events are important and relevant and what events are merely distractions the events themselves do not indicate. They are not even symptoms in the sense in which the patient's narrative is a clue for the physician.

If the executive lets the flow of events determine what he does, what he works on, and what he takes seriously, he will fritter himself away "operating." He may be an excellent person. But he is certain to waste his knowledge and ability and to throw away what little effectiveness he might have achieved. What the executive needs are criteria that enable him to work on the truly important, that is, on contributions and results, even though the criteria are not found in the flow of events.

3. The third reality pushing the executive toward ineffectiveness is that he is within an *organization*. This means that he is effective only if and when other people make use of what he contributes. Organization is a means of multiplying the strength of an individual. It takes his knowledge and uses it as the resource, the motivation, and the vision of other knowledge workers. Knowledge workers are rarely in synch with each other, precisely because they are knowledge workers. Each has his or her own skill and concerns. One may be interested in tax accounting or in bacteriology, or in training and developing tomorrow's key administrators in the city government. But the worker next door is interested in the finer points of cost accounting, in hospital economics, or in the legalities of the city charter. Each has to be able to use what the other produces.

Usually the people who are most important to the effectiveness of an executive are not people over whom he has direct control. They are people in other areas, people who in terms of organization, are "sideways." Or they are his superiors. Unless the executive can reach those people, can make his contribution effective for them and in their work, he has no effectiveness at all.

4. Finally, the executive is *within* an organization.

Every executive, whether his organization is a business or a research laboratory, a government agency, a large university, or the air force, sees the inside—the organization—as close and immediate reality. He sees the outside only through thick and distorting lenses, if at all. What goes on outside is usually not even known firsthand. It is received through an organizational filter of reports, that is, in an already predigested and highly abstract form that imposes organizational criteria of relevance on the outside reality.

Specifically, there are no results within the organization. All the results are on the outside. The only business results, for instance, are produced by a customer who converts the costs and efforts of the business into revenues and profits through his willingness to exchange his purchasing power for the products or services of the business.

What happens inside any organization is effort and cost. To speak of "profit centers" in a business as we are wont to do is polite euphemism. There are only effort centers. The less an organization has to do to produce results, the better it does its job. That it takes one hundred thousand employees to produce the automobiles or the steel the market wants is essentially a gross engineering imperfection. The fewer people, the smaller, the less activity inside, the more nearly perfect is the organization in terms of its only reason for existence: the service to the environment.

An organization is not, like an animal, an end in itself, and successful by the mere act of perpetuating the species. An organization is an organ of society and fulfills itself by the contribution it makes to the outside environment. And yet the bigger and apparently more successful an organization gets to be, the more will inside events tend to engage the interests, the energies, and the abilities of the executive to the exclusion of his real tasks and his real effectiveness in the outside.

This danger is being aggravated today by the advent of the computer and of the new information technology. The computer, being a mechanical moron, can handle only quantifiable data. These it can handle with speed, accuracy, and precision. It will, therefore, grind out hitherto unobtainable quantified information in large volume. One can, however, by and large quantify only what goes on inside an organization—costs and production figures, patient statistics in the hospital, or training reports. The relevant outside events are rarely available in quantifiable form until it is much too late to do anything about them.

This is not because our information-gathering capacity in respect to the outside events lags behind the technical abilities of the computer. If that was the only thing to worry about, we would just have to increase statistical efforts—and the computer itself could greatly help us to overcome this mechanical limitation. The problem is rather that the important and relevant outside events are often qualitative and not capable of quantification. They are not yet "facts." For a fact, after all, is an event that somebody has defined, has classified, and, above all, has endowed with relevance. To be able to quantify, one has to have a concept first. One first has to abstract from the infinite welter of phenomena a specific aspect that one then can name and finally count.

The truly important events on the outside are not the trends. They are changes in the trends. These determine ultimately success or failure of an organization and its efforts. Such changes, however, have to be perceived; they cannot be counted, defined, or classified. The classifications still produce the expected figures—as they did for the Edsel. But the figures no longer correspond to actual behavior.

The computer is a logic machine, and that is its strength—but also its limitation. The important events on the outside cannot be reported in the kind of form a computer (or any other logic system) could possibly handle. Man, however, while not particularly logical is perceptive—and that is his strength.

The danger is that executives will become contemptuous of information and stimuli that cannot be reduced to computer logic and computer language. Executives may become blind to everything that is perception (i.e., event) rather than fact (i.e., after the event). The tremendous amount of computer information may thus shut out access to reality.

Eventually the computer—potentially by far the most useful management tool—should make executives aware of their insulation and free them for more time on the outside. In the short run, however, there is danger of acute "computeritis." It is a serious affliction.

The computer only makes visible a condition that existed before it. Executives of necessity live and work within an organization.

Unless they make conscious efforts to perceive the outside, the inside may blind them to the true reality.

These four realities the executive cannot change. They are necessary conditions of his existence. But he must therefore assume that he will be ineffectual unless he makes special efforts to learn to be effective.

The Promise of Effectiveness

Increasing effectiveness may well be the only area where we can hope significantly to raise the level of the knowledge worker's performance, achievement, and satisfaction.

We certainly could use people of much greater abilities in many places. We could use people of broader knowledge. I submit, however, that in these two areas, not too much can be expected from further efforts. We may be getting to the point where we are already attempting to do the inherently impossible or at least the inherently unprofitable. But we are not going to breed a new race of supermen. We will have to run our organizations with men and women as they are.

The books on manager development, for instance, envisage truly a "man for all seasons" in their picture of "the manager of tomorrow." A senior executive, we are told, should have extraordinary abilities as an analyst and as a decision-maker. He or she should be good at working with people and at understanding organization and power relations, be good at mathematics, and have artistic insights and creative imagination. What seems to be wanted is universal genius, and universal genius has always been in scarce supply. The experience of the human race indicates strongly that the only person in abundant supply is the universal incompetent. We will therefore have to staff our organizations with people who at best excel in one of these abilities. And then they are more than likely to lack any but the most modest endowment in the others.

We will have to learn to build organizations in such a manner that anybody who has strength in one important area is capable of putting it to work. But we cannot expect to get the performance we need by raising our standards for abilities, let alone by hoping for the universally gifted person. We will have to extend the range of human beings through the tools they have to work with rather than through a sudden quantum jump in human ability.

The same, more or less, applies to knowledge. However badly we may need people of more and better knowledge, the effort needed to make the major improvement may well be greater than any possible, let alone any probable, return.

When "operations research" first came in, several of the brilliant young practitioners published their prescription for the operations researcher of tomorrow. They always came out asking for a polymath knowing everything and capable of doing superior and original work in every area of human knowledge. According to one of these studies, operations researchers need to have advanced knowledge in sixty-two or so major scientific and humanistic disciplines. If such a person could be found, he would, I am afraid, be totally wasted on studies of inventory levels or on the programming of production schedules.

Much less ambitious programs for manager development call for high knowledge in such a host of divergent skills as accounting and personnel, marketing, pricing and economic analysis, the behavioral sciences such as psychology, and the natural sciences from physics to biology and geology. And we surely need people who understand the dynamics of modern technology, the complexity of the modern world economy, and the labyrinth of modern government.

Every one of these is a big area, is indeed too big even for those who work on nothing else. The scholars tend to specialize in fairly small segments of each of these fields and do not pretend to have more than a journeyman's knowledge of the field itself.

I am not saying that one need not try to understand the fundamentals of every one of these areas.

One of the weaknesses of young, highly educated people today—whether in business, medicine, or government—is that they are satisfied to be versed in one narrow specialty and affect a contempt for the other areas. One need not know in detail what to do with "human relations" as an accountant, or how to promote a new branded product if an engineer. But one has a responsibility to know at least what these areas are about, why they are around, and what they are trying to do. One need not know psychiatry to be a good urologist. But one had better know what psychiatry is all about. One need not be an international lawyer to do a good job in the Department of Agriculture. But one had better know enough about international politics not to do international damage through a parochial farm policy.

This, however, is something very different from the universal expert, who is as unlikely to occur as is the universal genius. Instead we will have to learn how to make better use of people who are good in any one of these areas. But this means increasing effectiveness. If one cannot increase the supply of a resource, one must increase its yield. And effectiveness is the one tool to make the resources of ability and knowledge yield more and better results.

Effectiveness thus deserves high priority because of the needs of organization. It deserves even greater priority as the tool of the executive and as his access to achievement and performance.

Can Effectiveness Be Learned?

If effectiveness were a gift people were born with, the way they are born with a gift for music or an eye for painting, we would be in bad shape. For we know that only a small minority is born with great gifts in any one of these areas. We would therefore be reduced to trying to spot people with high potential of effectiveness early and to train them as best we know to develop their talent. But we could hardly hope to find enough people for the executive tasks of modern society this way. Indeed, if effectiveness were a gift, our present civilization would be highly vulnerable, if not untenable. As a civilization of large organizations it is dependent on a large supply of people capable of being executives with a modicum of effectiveness.

If effectiveness can be learned, however, the questions arise: What does it consist in? What does one have to learn? Of what kind is the learning? Is it a knowledge—and knowledge one learns in systematic form and through concepts? Is it a skill that one learns as an apprentice? Or is it a practice that one learns through doing the same elementary things over and over again?

I have been asking those questions for a good many years. As a consultant, I work with executives in many organizations. Effectiveness is crucial to me in two ways. First, a consultant who by definition has no authority other than that of knowledge must himself be effective—or else he is nothing. Second, the most effective consultant depends on people within the client organization to get anything done. Their effectiveness therefore determines in the last analysis whether a consultant contributes and achieves results, or whether he is pure "cost center" or at best a court jester.

I soon learned that there is no "effective personality." The effective people I have seen differ widely in their temperaments and their abilities, in what they do and how they do it, in their personalities, their knowledge, their interests—in fact in almost everything that distinguishes human beings. All they have in common is the ability to get the right things done.

Among the effective people I have known and worked with, there are extroverts and aloof, retiring men, some even morbidly shy. Some are eccentrics, others painfully correct conformists. Some are fat and some are lean. Some are worriers; some are relaxed. Some drink quite heavily; others are total abstainers. Some are men of great charm and warmth; some have no more personality than a frozen mackerel. There are a few men among them who would answer to the popular conception of a "leader." But equally there are colorless men who would attract no attention in a crowd. Some are scholars and serious students, others almost unlettered. Some have broad interests; others know nothing except their own narrow area and care for little else. Some of the men are self-centered, if not indeed selfish. But there are also some who are generous of heart and mind. There are men who live only for their work and others whose main interests lie outside—in community work, in their church, in the study of Chinese poetry, or in modern music. Among the effective people I have met, there are people who use logic and analysis and others who rely mainly on perception and intuition. There are men who make decisions easily and men who suffer agonies every time they have to move.

Effective people, in other words, differ as widely as physicians, high-school teachers, or violinists. They differ as widely as do ineffectual ones, are indeed indistinguishable from ineffectual people in type, personality, and talents.

What all these effective people have in common is the practices that make effective whatever they have and whatever they are. And these practices are the same, whether he or she works in a business or in a government agency, as hospital administrator, or as university dean.

But whenever I have found one, no matter how great the intelligence, the industry, the imagination, or the knowledge, who fails to observe these practices, I have also found one deficient in effectiveness.

Effectiveness, in other words, is a habit; that is, a complex of practices. And practices can always be learned. Practices are simple, deceptively so; even a seven-year-old has no difficulty in understanding a practice. But practices are always exceedingly hard to do well. They have to be acquired, as we all learn the multiplication table; that is, repeated ad nauscam until "6 × 6 = 36" has become unthinking, conditioned reflex, and firmly ingrained habit. Practices one learns by practicing and practicing and practicing again.

To every practice applies what my old piano teacher said to me in exasperation when I was a small boy. "You will never play Mozart the way Arthur Schnabel does, but there is no reason in the world why you should not play your scales the way he does." What the piano teacher forgot to add—probably because it was so obvious to her—is that even the great pianists could not play Mozart as they do unless they practiced their scales and kept on practicing them.

There is, in other words, no reason why anyone with normal endowment should not acquire competence in any practice. Mastery might well elude him; for that one might need special talents. But what is needed in effectiveness is competence. What is needed are "the scales."

14.

FOCUS ON CONTRIBUTION

The effective person focuses on contribution. He looks up from his work and outward toward goals. He asks, "What can I contribute that will significantly affect the performance and the results of the institution I serve?" His stress is on responsibility.

The focus on contribution is the key to effectiveness: in one's own work (its content, its level, its standards, and its impacts) in one's relations with others (with superiors, associates, subordinates), and in the use of the tools of the executive such as meetings or reports.

The great majority of people tend to focus downward. They are occupied with efforts rather than with results. They worry over what the organization and their superiors "owe" them and should do for them. And they are conscious above all of the authority they "should have." As a result, they render themselves ineffectual.

The head of one of the large management consulting firms always starts an assignment with a new client by spending a few days visiting the senior executives of the client organization one by one. After he has chatted with them about the assignment and the client organization, its history and its people, he asks (though rarely, of course, in these words), "And what do *you* do that justifies your being on the payroll?" The great majority, he reports, answer, "I run the accounting department," or "I am in charge of the sales force." Indeed, not uncommonly the answer is, "I have eight hundred and fifty people working under me." Only a few say, "It's my job to give our managers the information they need to make the right decisions," or "I am responsible for finding out what products the customer will want tomorrow," or "I have to think through and prepare the decisions the president will have to face tomorrow."

The man who focuses on efforts and who stresses his downward authority is a subordinate no matter how exalted his title and rank. But the man who focuses on contribution and who takes responsibility for results, no matter how junior, is in the most literal sense of the phrase, "top management." He holds himself accountable for the performance of the whole.

Own Commitment

The focus on contribution turns one's attention away from his own specialty, his own narrow skills, his own department, and toward the performance of the whole. It turns his attention to the outside, the only place where there are results. He is likely to have to think through what relationships his skills, his specialty, his function, or his department have to the entire organization and *its* purpose. He therefore will also come to think in terms of the customer, the client, or the patient, who is the ultimate reason for whatever the organization produces, whether it be economic goods, governmental policies, or health services. As a result, what he does and how he does it will be materially different.

A large scientific agency of the U.S. government found this out a few years ago. The old director of publications retired. He had been with the agency since its inception in the 1930s and was neither scientist nor trained writer. The publications that he turned out were often criticized for lacking professional polish. He was replaced by an accomplished science writer. The publications immediately took on a highly professional look. But the scientific community for whom these publications were intended stopped reading them. A highly respected university scientist, who had for many years worked closely with the agency, finally told the administrator, "The former director was writing *for* us; your new man writes *at* us."

The old director had asked the question, "What can I contribute to the results of this agency?" His answer was, "I can interest the young scientists on the outside in our work, can make them want to come to work for us." He therefore stressed major problems, major decisions, and even major controversies inside the agency. This had brought him more than once into head-on collision with the administrator. But the old man had stood by his guns. "The test of our publications is not whether we like them; the test is how many young scientists apply to us for jobs and how good they are," he said.

To ask, "What can I contribute?" is to look for the unused potential in the job. And what is considered excellent performance in a good many positions is often but a pale shadow of the job's full potential of contribution.

Knowledge workers who do not ask themselves, "What can I contribute?" are not only likely to aim too low, they are likely to aim at the wrong things. Above all, they may define their contribution too narrowly.

"Contribution" may mean different things. For every organization needs performance in three major areas: direct results, building of values and their reaffirmation, and building and developing people for tomorrow. If deprived of performance in any one of these areas, it will decay and die. All three therefore have to be built into the contribution of every knowledge worker. But their relative importance varies greatly with the personality and the position of the knowledge worker as well as with the needs of the organization.

The direct results of an organization are clearly visible, as a rule. In a business, they are economic results such as sales and profits; in a hospital, they are patient care; and so on. But sometimes even direct results are not totally unambiguous. And when there is confusion as to what they should be, there are no results.

Direct results always come first. In the care and feeding of an organization, they play the role calories play in the nutrition of the human body. But any organization also needs a commitment to values and their constant reaffirmation, as a human body needs vitamins and minerals. There has to be something "this organization stands for," or else it degenerates into disorganization, confusion, and paralysis. In a business, the value commitment may be to technical leadership or (as in Sears, Roebuck) to finding the right goods and services for the American family and to procuring them at the lowest price and the best quality.

Value commitments, like results, are not unambiguous.

The U.S. Department of Agriculture has for many years been torn between two fundamentally incompatible value commitments—one to agricultural productivity and one to the "family farm" as the "backbone of the nation." The former has been pushing the country toward industrial agriculture, highly mechanical, highly industrialized, and essentially a large-scale commercial business. The latter has called for nostalgia supporting a nonproducing rural proletariat. But because farm policy—at least until very recently—has wavered between two different value commitments, all it has really succeeded in doing has been to spend prodigious amounts of money.

Finally, organization is, to a large extent, a means of overcoming the limitations mortality sets to what any one person can contribute. An organization that is not capable of perpetuating itself has failed. An organization therefore has to provide today the men and women who can run it tomorrow. It has to renew its human capital. It should steadily upgrade its human resources. The next generation should take for granted what the hard work and dedication of this generation has accomplished. They should then, standing on the shoulders of their predecessors, establish a new "high" as the baseline for the generation after them.

An organization that just perpetuates today's level of vision, excellence, and accomplishment has lost the capacity to adapt. And since the one and only thing certain in human affairs is change, it will not be capable of survival in a changed tomorrow.

Focus on contribution by itself is a powerful force in developing people. People adjust to the level of the demands made on them. One who sets his sights on contribution raises the sights and standards of everyone with whom he works.

A new hospital administrator, holding his first staff meeting, thought that a rather difficult matter had been settled to everyone's satisfaction when one of the participants suddenly asked, "Would this have satisfied Nurse Bryan?" At once the argument started all over and did not subside until a new and much more ambitious solution to the problem had been hammered out.

Nurse Bryan, the administrator learned, had been a long-serving nurse at the hospital. She was not particularly distinguished, had not in fact ever been a supervisor. But whenever a decision on a patient's care came up on her floor, Nurse Bryan would ask, "Are we doing the best we can do to help this patient?" Patients on Nurse Bryan's floor did better and recovered faster. Gradually over the years, the whole hospital had learned to adopt what came to be known as Nurse Bryan's Rule; had learned, in other words, to ask, "Are we really making the best contribution to the purpose of this hospital?"

Though Nurse Bryan herself had retired almost ten years earlier, the standards she had set still made demands on people who in terms of training and position were her superiors.

Commitment to contribution is commitment to responsible effectiveness. Without it, a person shortchanges himself, deprives his organization, and cheats the people he works with.

The most common cause of failure is inability or unwillingness to change with the demands of a new position. The knowledge worker who keeps on doing what he has done successfully before he moved is almost bound to fail. Not only do the results change to which his contribution ought to direct itself. The relative importance among the three dimensions of performance changes. The knowledge worker who fails to understand this will suddenly do the wrong things the wrong way—even though he does exactly what in his old job had been the right things done the right way.

Contribution of Knowledges

For the knowledge worker to focus on contribution is particularly important. This alone can enable him to contribute at all.

Knowledge workers do not produce a "thing." They produce ideas, information, concepts. The knowledge worker, moreover, is usually a specialist. In fact, he can, as a rule, be effective only if he has learned to do one thing very well, that is, if he has specialized. By itself, however, a specialty is a fragment and sterile. Its output has to be put together with the output of other specialists before it can produce results.

The task is not to breed generalists. It is to enable the specialist to make himself and his specialty effective. This means that he must think through who is to use his output and what the user needs to know and to understand to be able to make productive the fragment the specialist produces.

The person of knowledge has always been expected to take responsibility for being understood. It is barbarian arrogance to assume that the layman can or should make the effort to understand the specialist, and that it is enough if the person of knowledge talks to a handful of fellow experts who are his peers. Even in the university or in the research laboratory, this attitude—alas, only too common today—condemns the expert to uselessness and converts his knowledge from learning into pedantry. If a person wants to be an executive—that is, if he wants to be considered responsible for his contribution—he has to concern himself with the usability of his "product"—that is, his knowledge.

Effective knowledge workers know this. For they are almost imperceptibly led by their upward orientation into finding out what the other fellow needs, what the other fellow sees, and what the other fellow understands. Effective people find themselves asking other people in the organization, their superiors, their subordinates, but above all, their colleagues in other areas, "What contribution from me do you require to make *your* contribution to the organization? When do you need this, how do you need it, and in what form?"

The person who takes responsibility for his contribution will relate his narrow area to a genuine whole. He may never himself be able to integrate a number of knowledge areas into one. But he soon realizes that he has to learn enough of the needs, the directions, the limitations, and the perceptions of others to enable them to use his own work. Even if this does not make him appreciate the richness and the excitement of diversity, it will give him immunity against the arrogance of the learned—that degenerative disease that destroys knowledge and deprives it of beauty and effectiveness.

The Right Human Relations

Knowledge workers in an organization do not have good human relations because they have a "talent for people." They have good human relations because they focus on contribution in their own work and in their relationships with others. As a result, their relationships are productive—and this is the only valid definition of "good human relations." Warm feelings and pleasant words are meaningless, are indeed a false front for wretched attitudes, if there is no achievement in what is, after all, a work-focused and task-focused relationship. On the other hand, an occasional rough word will not disturb a relationship that produces results and accomplishments for all concerned.

The focus on contribution by itself supplies the four basic requirements of effective human relations:

- Communications
- Teamwork

➤ Self-development

➤ Development of others

1. Communications have been in the center of managerial attention these last twenty years or more. In business, in public administration, in armed services, in hospitals, in other words in all the major institutions of modern society, there has been great concern with communications.

Results to date have been meager. Communications are by and large just as poor today as they were twenty or thirty years ago when we first became aware of the need for, and lack of, adequate communications in the modern organization. But we are beginning to understand why this massive communications effort cannot produce results.

We have been working at communications downward from management to the employees, from the superior to the subordinate. But communications are practically impossible if they are based on the downward relationship. This much we have learned from our work in perception and communications theory. The harder the superior tries to say something to his subordinate, the more likely is it that the subordinate will *mis*hear. He will hear what he expects to hear rather than what is being said.

But knowledge workers who take responsibility for contribution in their own work will as a rule demand that their subordinates take responsibility, too. They will tend to ask their subordinates: "What are the contributions for which this organization and I, your superior, should hold you accountable? What should we expect of you? What is the best utilization of your knowledge and your ability?" And then communication becomes possible, becomes indeed easy.

Once the subordinate has thought through what contribution should be expected of him or her, the superior has, of course, both the right and the responsibility to judge the validity of the proposed contribution. According to all our experience, the objectives set by subordinates for themselves are almost never what the superior thought they should be. The subordinates or juniors, in other words, do see reality quite differently. And the more capable they are, the more willing to take responsibility, the more will their perception of reality and of its objective opportunities and needs differ from the view of their superior or of the organization. But any discrepancy between their conclusions and what their superior expected will stand out strongly.

Who is right in such a difference is not as a rule important. For effective communication in meaningful terms has already been established.

2. The focus on contribution leads to communications sideways and thereby makes teamwork possible.

The question, "Who has to use my output for it to become effective?" immediately shows up the importance of people who are not in line of authority, either upward or downward, from and to the individual executive. It underlines what is the reality of a knowledge organization: the effective work is actually done in and by teams of people of diverse knowledges and skills. Those people have to work together voluntarily and according to the logic of the situation and the demands of the task, rather than according to a formal jurisdictional structure.

In a hospital, for instance—perhaps the most complex of the modern knowledge organizations—nurses, dieticians, physical therapists, medical and X-ray technicians, pharmacologists, pathologists, and a host of other health-service professionals have to work on and with the same patient, with a minimum of conscious command or control by anyone. And yet they have to work together for a common end and in line with a general plan of action: the doctor's prescription for treatment. In terms of organizational structure, each of these health-service professionals reports to his own chief. Each operates in terms of his own highly specialized field of knowledge; that is, as a "professional." But each has to keep all the others informed according to the specific situation, the condition, and the need of an individual patient. Otherwise, their efforts are more likely to do harm than good.

In a hospital in which the focus on contribution has become

ingrained habit, there is almost no difficulty in achieving such teamwork. In other hospitals this sideways communication, this spontaneous self-organization into the right task-focused teams, does not occur despite frantic efforts to obtain communications and coordination through all kinds of committees, staff conferences, bulletins, sermons, and the like.

3. Individual self-development in large measure depends on the focus on contributions.

The man who asks of himself, What is the most important contribution I can make to the performance of this organization? asks in effect, What self-development do I need? What knowledge and skill do I have to acquire to make the contribution I should be making? What strengths do I have to put to work? What standards do I have to set myself?

4. The executive who focuses on contribution also stimulates others to develop themselves, whether they are subordinates, colleagues, or superiors. He sets standards that are not personal but grounded in the requirements of the task. At the same time, they are demands for excellence. For they are demands for high aspiration, for ambitious goals, and for work of great impact.

We know very little about self-development. But we do know one thing: people in general, and knowledge workers in particular, grow according to the demands they make on themselves. They grow according to what they consider to be achievement and attainment. If they demand little of themselves, they will remain stunted. If they demand a good deal of themselves, they will grow to giant stature—without any more effort than is expended by the nonachievers.

15.

KNOW YOUR STRENGTHS
AND VALUES

More and more people in the workforce—and most knowledge workers—will have to *manage themselves.* They will have to place themselves where they can make the greatest contribution; they will have to learn to develop themselves. They will have to learn to stay young and mentally alive during a fifty-year working life. They will have to learn how and when to change what they do, how they do it, and when they do it.

Knowledge workers are likely to outlive their employing organization. Even if knowledge workers postpone entry into the labor force as long as possible—if, for instance, they stay in school till their late twenties to get a doctorate—they are likely, with present life expectancies in the developed countries, to live into their eighties. And they are likely to have to keep working, if only part-time, until they are around seventy-five or older. The average working life, in other words, is likely to be fifty years, especially for knowledge workers. But the average life expectancy of a successful business is only thirty years—and in a period of great turbulence such as the one we are living in, it is unlikely to be even that long. Increasingly, therefore, workers, and especially knowledge workers, will outlive any one employer, and will have to be prepared for more than one job, more than one assignment, more than one career.

What Are My Strengths?

Most people think they know what they are good at. They are usually wrong. People know what they are *not* good at more often—and even there people are more often wrong than right. And yet, one can only perform with one's strengths. One cannot build performance on weaknesses, let alone on something one cannot do at all.

For the great majority of people, to know their strengths was irrelevant only a few decades ago. One was born into a job and into a line of work. The peasant's son became a peasant. If he was not good at being a peasant, he failed. The artisan's son was similarly going to be an artisan, and so on. But now people have choices. They therefore have to know their strengths so that they can know where they belong.

There is only one way to find out: *the feedback analysis.* Whenever one makes a key decision, and whenever one does a key action, one writes down what one expects will happen. And nine months or twelve months later, one then feeds back from results to expectations. I have been doing this for some fifteen to twenty years now. And every time I do it, I am surprised. And so is every one who has ever done this.

Within a fairly short period of time, maybe two or three years, this simple procedure will tell people first where their strengths are—and this is probably the most important thing to know about oneself. It will show them what they do or fail to do that deprives them of the full yield from their strengths. It will show them where they are not particularly competent. And finally it will show them where they have no strengths and cannot perform.

Several *action conclusions* follow from the feedback analysis.

The first, and most important, conclusion: *Concentrate on your strengths.* Place yourself where your strengths can produce performance and results.

Second: *Work on improving your strengths.* The feedback analysis rapidly shows where a person needs to improve skills or has to acquire new knowledge. It will show where existing skills and knowledge are no longer adequate and have to be updated. It will also show the gaps in one's knowledge.

And one can usually acquire enough of any skill or knowledge not to be incompetent in it.

Of particular importance is the third conclusion: *Identify where intellectual arrogance causes disabling ignorance.* The feedback analysis soon identifies these areas. Far too many people—and especially people with high knowledge in one area—are contemptuous of knowledge in other areas or believe that being "bright" is a substitute for knowing. And then the feedback analysis soon shows that a main reason for poor performance is the result of simply not knowing enough, or the result of being contemptuous of knowledge outside one's own specialty.

An equally important action conclusion is: *Remedy your bad habits*—things you do or fail to do that inhibit your effectiveness and performance. They quickly show up in the feedback analysis.

But the analysis may also show that a person fails to obtain results because he or she lacks *manners.* Bright people—especially bright young people—often do not understand that manners are the "lubricating oil" of an organization.

The next action conclusion from the feedback analysis involves what *not* to do.

Feeding back from results to expectations soon shows where a person should not try to do anything at all. It shows the areas in which a person lacks the minimum endowment needed—and there are always many such areas for any person. Not enough people have even one first-rate skill or knowledge area, but all of us have an infinite number of areas in which we have no talent, no skill, and little chance to become even mediocre. And in these areas a person—and especially a knowledge worker—should not take on work, jobs, assignments.

The final action conclusion is: *Waste as little effort as possible on improving areas of low competence.* Concentration should be on areas of high competence and high skill. It takes far more energy and far more work to improve from incompetence to low mediocrity than it takes to improve from first-rate performance to excellence. And yet most people—and equally most teachers and most organizations—try to concentrate on making an incompetent person into a low mediocrity. The energy and resources—and time—should instead go into making a competent person into a star performer.

How Do I Perform?

How do I perform? is as important a question—and especially for knowledge workers—as What are my strengths?

In fact, it may be an even more important question. Amazingly few people know *how* they get things done. Most of us do not even know that different people work and perform differently. They therefore work in ways that are not their ways—and that almost guarantees nonperformance.

Like one's strengths, how one performs is *individual*. It is *personality*. Whether personality be "nature" or "nurture," it surely is formed long before the person goes to work. And *how* a person performs is a "given," just as *what* a person is good at or not good at is a "given." It can be modified, but it is unlikely to be changed. And just as people achieve results by doing *what* they are good at, people also achieve results by performing *how* they perform.

The feedback analysis may indicate that there is something amiss in how one performs. But rarely does it identify the cause. It is, however, normally not too difficult to find out. It takes a few years of work experience. And then one can ask—and quickly answer—*how* one performs. For a few common personality traits usually determine how one achieves results.

The first thing to know about how one performs is whether one is a reader or a listener. Yet very few people even know that there are readers and there are listeners, and that very few people are both. Even fewer know which of the two they themselves are. But few things are as damaging as not to know whether one is a reader or a listener.

The second thing to know about how one performs is to know how one *learns*. There things may be even worse than they are in respect to readers and listeners. For schools everywhere are organized on the totally erroneous assumption that there is one right way to learn, and that it is the same way for everybody.

Here is an example of one of the different ways in which people learn.

Beethoven left behind an enormous number of notebooks. Yet he himself said that he never looked at a notebook when he actually wrote his compositions. When asked, "Why then, do you keep a notebook?" he is reported to have answered, "If I don't write it down immediately, I forget it right away. If I put it into a notebook, I never forget it, and I never have to look it up again."

There are probably half a dozen different ways to learn. There are people who learn by taking copious notes—the way Beethoven did. But Alfred Sloan never took a note in a meeting. There are people who learn by hearing themselves talk. There are people who learn by writing. There are people who learn by doing. And in an informal survey I once took of professors in American universities who successfully publish scholarly books of wide appeal, I was told again and again, "To hear myself talk is the reason why I teach; because then I can write."

Actually, of all the important pieces of self-knowledge, this is one of the easiest to acquire. When I ask people, "How do you learn?" most of them know it. But when I then ask, "Do you act on this knowledge?" few reply that they do. And yet to act on this knowledge is the key to performance—or rather *not* to act on this knowledge is to condemn oneself to nonperformance.

How do I perform? and How do I learn? are the most important first questions to ask. But they are by no means the only ones. To manage oneself, one has to ask, Do I work well with people, or am I a loner? And if one finds out that one works well with people, one asks, In what relationship do I work well with people?

Some people work best as team members. Some people work exceedingly well as coaches and mentors, and some people are simply incompetent to be mentors.

Another important thing to know about how one performs is whether one performs well *under stress,* or whether one needs a highly structured and predictable environment. Another trait: Does one work best as a minnow in a big organization, or as a big fish in a small organization? Few people work well in both ways. Again and again people who have been very successful in a large organization— for example, the General Electric Company or Citibank—flounder miserably when they move into a small organization. And again and again people who perform brilliantly in a small organization flounder miserably when they take a job with a big organization.

Another crucial question: Do I produce results as a decision-maker or as an adviser? A great many people perform best as advisers, but cannot take the burden and pressure of the decision. A good many people, by contrast, need an adviser to force themselves to think, but then they can make the decision and act on it with speed, self-confidence, and courage.

This is a reason, by the way, why the number-two person in an organization often fails when promoted into the top spot. The top spot requires a decision-maker. Strong decision-makers in the top spot often put somebody whom they trust into the number-two spot as their adviser—and in that position that person is outstanding. But when then promoted into the number-one spot, the person fails. He or she knows what the decision should be but cannot take decision-*making* responsibility.

The *action conclusion:* Again, *do not try to change yourself*—it is unlikely to be successful. But work, and hard, to improve the way you perform. And try not to do work of any kind in a way in which you do not perform, or perform poorly.

What Are My Values?

To be able to manage oneself, one finally has to know, What are my values?

Organizations have to have values. But so do people. To be effective in an organization, one's own values must be compatible with the organization's values. They do not need to be the same. But they must be close enough so that they can coexist. Otherwise, the person will be frustrated, but also the person will not produce results.

There rarely is a conflict between a person's strengths and the way that person performs. The two are complementary. But there is sometimes a conflict between a person's values and the same person's strengths. What one does well—even very well—and successfully may not fit with one's value system. It may not appear to that person as making a contribution and as something to which to devote one's life (or even a substantial portion thereof).

If I may inject a personal note: I too, many years ago, had to decide between what I was doing well and successfully, and my values. I was doing extremely well as a young investment banker in London in the mid-1930s; it clearly fitted my strengths. Yet I did not see myself making a contribution as an asset manager of any kind. People, I realized, were my values. And I saw no point in being the richest man in the cemetery. Those were the darkest days of the Great Depression; I had no money, no job, and no prospects. But I quit—and it was the right thing.

Values, in other words, are and should be the ultimate test.

Where Do I Belong?

The answers to the three questions, What are my strengths? How do I perform? and What are my values? should enable the individual, and especially the individual knowledge worker, to decide where he or she belongs.

This is not a decision that most people can or should make at the beginning of their careers. But most people, and especially highly gifted people, do not really know where they belong till they are well past their mid-twenties. By that time, however, they should know where their strengths are. They should know how they perform. And they should know what their values are.

And then they can and should decide where they belong. Or rather, they should be able to decide where they do *not belong*. The person who has learned that he or she does not really perform in a big organization should have learned to say no when offered a position in a big organization. The person who has learned that he or she is not a decision-maker should have learned to say no when offered a decision-making assignment.

But also knowing the answer to these three questions enables people to say to an opportunity, to an offer, to an assignment, "Yes, I'll do that. But this is the way *I* should be doing it. This is the way it should be structured. This is the way my relationships should be. These are the kinds of results you should expect from me, and in this time frame, because *this is who I am*."

Successful careers are not "planned." They are the careers of people who are prepared for the opportunity because they know their strengths, the way they work, and their values. For knowing where one belongs makes ordinary people—hardworking, competent, but mediocre otherwise—into outstanding performers.

16.

KNOW YOUR TIME

Most discussions of the knowledge worker's task start with the advice to plan one's work. This sounds eminently plausible. The only thing wrong with it is that it rarely works. The plans always remain on paper, always remain good intentions. They seldom turn into achievement.

Effective knowledge workers, in my observation, do not start with their tasks. They start with their time. And they do not start out with planning. They start by finding out where their time actually goes. Then they attempt to manage their time and to cut back unproductive demands on their time. Finally they consolidate their "discretionary" time into the largest possible continuing units. This three-step process:

> ➤ Recording time
> ➤ Managing time
> ➤ Consolidating time

is the foundation of executive effectiveness.

Effective people know that time is the limiting factor. The output limits of any process are set by the scarcest resource. In the process we call "accomplishment," that resource is time.

Time is also a unique resource. One cannot rent, hire, buy, or otherwise obtain more time.

The supply of time is totally inelastic. No matter how high the demand, the supply will not increase. There is no price for it and no marginal utility curve for it. Moreover, time is totally perishable and cannot be stored. Yesterday's time is gone forever and will never come back. Time is, therefore, always in exceedingly short supply.

Time is totally irreplaceable. Within limits we can substitute one resource for another, copper for aluminum, for instance. We can substitute capital for human labor. We can use more knowledge or more brawn. But there is no substitute for time.

Everything requires time. It is the one truly universal condition. All work takes place in time and uses up time. Yet most people take for granted this unique, irreplaceable, and necessary resource. Nothing else, perhaps, distinguishes effective executives as much as their tender loving care of time.

Man is ill-equipped to manage his time. Even in total darkness, most people retain their sense of space. But even with the lights on, a few hours in a sealed room render most people incapable of estimating how much time has elapsed. They are as likely to underrate grossly the time spent in the room as to overrate it grossly.

If we rely on our memory, therefore, we do not know how time has been spent.

I sometimes ask executives who pride themselves on their memory to put down their guess as to how they spend their own time. Then I lock these guesses away for a few weeks or months. In the meantime, the executives run an actual time record on themselves. There is never much resemblance between the way these people thought they used their time and their actual records.

One company chairman was absolutely certain that he divided his time roughly into three parts. One third he thought he was spending with his senior men. One-third he thought he spent with his important customers. And one-third he thought was devoted to community activities. The actual record of his activities over six weeks brought out clearly that he spent almost no time in any of these areas. These were the tasks on which he knew he *should* spend time—and therefore memory, obliging as usual, told him that they were the tasks on which he actually had spent his time. The record showed, however, that he spent most of his hours as a kind of dispatcher, keeping track of orders from customers he personally knew, and bothering the plant with telephone calls about them. Most of those orders were going through all right anyhow and his intervention could only delay them. But when his secretary first came in with the time record, he did not believe her. It took two or three more time logs to convince him that the record, rather than his memory, had to be trusted when it came to the use of time.

The effective person therefore knows that to manage his time, he first has to know where it actually goes.

The Time Demands

There are constant pressures toward unproductive and wasteful time-use. Any knowledge worker, whether he is a manager or not, has to spend a great deal of his time on things that do not contribute at all. Much is inevitably wasted. The higher up in the organization he is, the more demands on his time will the organization make.

The head of a large company once told me that in two years as chief executive officer he had eaten out every evening except on Christmas Day and New Year's Day. All the other dinners were "official" functions, each of which wasted several hours. Yet he saw no possible alternative. Whether the dinner honored an employee retiring after fifty years of service, or the governor of one of the states in which the company did business, the chief executive officer had to be there. Ceremony is one of his tasks. My friend had no illusions that these dinners contributed anything either to the company or to his own entertainment or self-development. Yet he had to be there and dine graciously.

Similar time-wasters abound in the life of every knowledge worker. When a company's best customer calls up, the sales manager cannot say, "I am busy." He has to listen, even though all the customer wants to talk about may be a bridge game the preceding Saturday or the chances of his daughter's getting into the right college. The hospital administrator has to attend the meetings of every one of his staff committees, or else the physicians, the nurses, the technicians, and other staff members will feel that they are being slighted. The government administrator had better pay attention when a congressman calls and wants some information he could, in less time, get out of the telephone book or the *World Almanac*. And so it goes all day long.

Nonmanagers are no better off. They too are bombarded with demands on their time that add little, if anything, to their productivity, and yet cannot be disregarded.

In every job, a large part of the time must therefore be wasted on things that, though they apparently have to be done, contribute nothing or little.

Yet most of the tasks of the knowledge worker require, for minimum effectiveness, a fairly large quantum of time. To spend in one stretch less than this minimum is sheer waste. One accomplishes nothing and has to begin all over again.

To write a report may, for instance, require six or eight hours, at least for the first draft. It is pointless to give seven hours to the task by spending fifteen minutes twice a day for three weeks. All one has at the end is blank paper with some doodles on it. But if one can lock the door, disconnect the telephone, and sit down to wrestle with the report for five or six hours without interruption, one has a good chance to come up with what I call a "zero draft"—the one before the first draft. From then on, one can indeed work in fairly small installments, can rewrite, correct, and edit section by section, paragraph by paragraph, sentence by sentence.

The same goes for an experiment. One simply has to have five to twelve hours in a single stretch to set up the apparatus and to do at least one completed run. Or one has to start all over again after an interruption.

To be effective, every knowledge worker, and especially every executive, therefore needs to be able to dispose of time in fairly large chunks. To have small dribs and drabs of time at his disposal will not be sufficient even if the total is an impressive number of hours.

This is particularly true with respect to time spent working with people, which is, of course, a central task in the work of the executive. People are time-consumers. And most people are time-wasters.

To spend a few minutes with people is simply not productive. If one wants to get anything across, one has to spend a fairly large minimum quantum of time. The knowledge worker who thinks that he can discuss the plans, direction, and performance of one of his subordinates in fifteen minutes—and many managers believe this—is just deceiving himself. If one wants to get to the point of having an impact, one needs probably at least an hour and usually much more. And if one has to establish a human relationship, one needs infinitely more time.

Relations with other knowledge workers are especially time-consuming. Whatever the reason—whether it is the absence of the barrier of class and authority between superior and subordinate in knowledge work, or whether he simply takes himself more seriously—the knowledge worker makes much greater time demands than the manual worker on his superior as well as on his associates. Moreover, because knowledge work cannot be measured the way manual work can, one cannot tell a knowledge worker in a few simple words whether he is doing the right job and how well he is doing it. One can say to a manual worker, "Our work standard calls for fifty pieces an hour, and you are only turning out forty-two." One has to sit down with a knowledge worker and think through with him what should be done and why, before one can even know whether he is doing a satisfactory job or not. And that is time-consuming.

Since the knowledge worker directs himself, he must understand what achievement is expected of him and why. He must also understand the work of the people who have to use his knowledge output. For this, he needs a good deal of information, discussion, instruction—all things that take time. And contrary to common belief, this time demand is made not only on his superior but equally on his colleagues.

The knowledge worker must be focused on the results and performance goals of the entire organization to have any results and performance at all. This means that he has to set aside time to direct his vision from his work to results, and from his specialty to the outside in which alone performance lies.

Wherever knowledge workers perform well in large organizations, senior executives take time out, on a regular schedule, to sit down with them, sometimes all the way down to green juniors, and ask, "What should we at the head of this organization know about your work? What do you want to tell me regarding this organization? Where do you see opportunities we do not exploit? Where do you see dangers to which we are still blind? And, all together, what do you want to know from me about the organization?"

This leisurely exchange is needed equally in a government agency and in a business, in a research lab and in an army staff. Without it, the knowledge people either lose enthusiasm and become time-servers, or they direct their energies toward their specialty and away from the opportunities and needs of the organization. But such a session takes a great deal of time, especially as it should be unhurried and relaxed. People must feel that "we have all the time in the world." This actually means that one gets a great deal done fast. But it means also that one has to make available a good deal of time in one chunk and without too much interruption.

Mixing personal relations and work relations is time-consuming. If hurried, it turns into friction. Yet any organization rests on this mixture. The more people are together, the more time will their sheer interaction take, the less time will be available to them for work, accomplishment, and results.

The larger the organization, therefore, the less actual time will the knowledge worker have and the more important it will be for him to know where his time goes and to manage the little time at his disposal.

The more people there are in an organization, the more often does a decision on people arise. But fast personnel decisions are likely to be wrong decisions. The time quantum of the good personnel decision is amazingly large. What the decision involves often becomes clear only when one has gone around the same track several times.

It is not the knowledge workers in the industrial countries of the world today who have a problem of spending their leisure time. On the contrary, they are working everywhere longer hours and have greater demands on their time to satisfy. And the time scarcity is bound to become worse rather than better.

One important reason for this is that a high standard of living presupposes an economy of innovation and change. But innovation and change make inordinate time demands on the executive. All one can think and do in a short time is to think what one already knows and to do as one has always done.

Time Diagnosis

That one has to record time before one can know where it goes and before, in turn, one can attempt to manage it we have realized for the best part of a century. That is, we have known this in respect to manual work, skilled and unskilled, since Scientific Management around 1900 began to record the time it takes for a specific piece of manual work to be done. Hardly any country is today so far behind in industrial methods as not to time systematically the operations of manual workers.

We have applied this knowledge to the work where time does not greatly matter, that is, where the difference between time-use and time-waste is primarily efficiency and costs. But we have not applied it to the work that matters increasingly, and that particularly has to cope with time: the work of the knowledge worker and especially of the executive. Here the difference between time-use and time-waste is effectiveness and results.

The first step toward effectiveness is therefore to record actual time-use. The specific method in which the record is put together need not concern us here. There are executives who keep such a time log themselves. Others, such as the company chairman just mentioned, have their secretaries do it for them. The important thing is that it gets done, and that the record is made in "real" time, that is, at the time of the event itself, rather than later on from memory.

A good many effective people keep such a log continually and look at it regularly every month. At a minimum, effective executives have the log run on themselves for three to four weeks at a stretch twice a year or so, on a regular schedule. After each such sample, they rethink and rework their schedule. But six months later they invariably find that they have "drifted" into wasting their time on trivia.

Time-use does improve with practice. But only constant efforts at managing time can prevent drifting. Systematic time management is therefore the next step. One has to find the nonproductive, time-wasting activities and get rid of them if one possibly can. This requires asking oneself a number of diagnostic questions.

1. First one tries to identify and eliminate the things that need not be done at all, the things that are purely waste of time without any results whatever. To find these time-wastes, one asks of *all* activities in the time records, What would happen if this were not done at all? And if the answer is, Nothing would happen, then obviously the conclusion is to stop doing it.

It is amazing how many things busy people are doing that never will be missed. There are, for instance, the countless speeches, dinners, committee meetings, and board meetings, which take an unconscionable toll of the time of busy people, which are rarely enjoyed by them or done well by them, but which are endured, year in and year out, as an Egyptian plague ordained from on high. Actually, all one has to do is to learn to say no if an activity contributes nothing to one's own organization, to oneself, or to the organization for which it is to be performed.

The chief executive mentioned above who had to dine out every night found, when he analyzed these dinners, that at least one-third would proceed just as well without anyone from the company's senior management. In fact, he found (somewhat to his chagrin) that his acceptance of a good many of these invitations was by no means welcome to his hosts. They had invited him as a polite gesture. But they had fully expected to be turned down and did not quite know what to do with him when he accepted.

I have yet to see a knowledge worker, regardless of rank or station, who could not consign something like a quarter of the demands on his time to the wastepaper basket without anybody's noticing their disappearance.

2. The next question is, Which of the activities on my time log could be done by somebody else just as well, if not better?

The dinner-eating company chairman found that any senior executive of the company would do for another third of the formal dinners—all the occasion demanded was the company's name on the guest list.

But I have never seen a knowledge worker confronted with his time record who did not rapidly acquire the habit of pushing at other people everything that he need not do personally. The first look at the time record makes it abundantly clear that there just is not time enough to do the things he himself considers important, himself wants to do, and is himself committed to doing. The only way he can get to the important things is by pushing on others anything at all that can be done by them.

"Delegation," as the term is customarily used, is a misnomer in this situation. But getting rid of anything that can be done by somebody else so that one does not have to delegate but can really get to one's own work—that is a major improvement in effectiveness.

3. A common cause of time-waste is largely under the executive's control and can be eliminated by him. That is the time of others he himself wastes.

There is no one symptom for this. But there is still a simple way to find out how and when it occurs. That is to ask other people. Effective people have learned to ask systematically and without coyness, "What do I do that wastes your time without contributing to your effectiveness?" To ask such a question, and to ask it without being afraid of the truth, is a mark of the effective executive.

The manner in which an executive does productive work may still be a major waste of somebody's else's time.

The senior financial executive of a large organization knew perfectly well that the meetings in his office wasted a lot of time. This man asked all his direct subordinates to every meeting, whatever the topic. As a result, the meetings were far too large. And because every participant felt that he had to show interest, everybody asked at least one question—most of them irrelevant—and the meetings stretched on endlessly. But the senior executive had not known, until he asked, that his subordinates too considered the meetings a waste of their time. Aware of the great importance everyone in the organization placed on status and on being "in the know," he had feared that anyone not invited would feel slighted and left out.

Now, however, he satisfies the status needs of his subordinates in a different manner. He sends out a printed form that reads: "I have asked [Messrs. Smith, Jones, and Robinson] to meet with me [Wednesday at 3] in [the fourth-floor conference room] to discuss [next year's capital-appropriations budget]. Please come if you think that you need the information or want to take part in the discussion. But you will in any event receive right away a full summary of the discussion and of any decisions reached, together with a request for your comments."

Where formerly a dozen people came and stayed all afternoon, three men and a secretary to take the minutes now get the matter over with in an hour or so. And no one feels left out.

Many knowledge workers know all about these unproductive

and unnecessary time demands; yet they are afraid to prune them. They are afraid to cut out something important by mistake. But such a mistake, if made, can be speedily corrected. If one prunes too harshly, one usually finds out fast enough.

But the best proof that the danger of overpruning is a bugaboo is the extraordinary effectiveness so often attained by severely ill or severely handicapped people.

A good example was Harry Hopkins, President Roosevelt's confidential adviser in World War II. A dying, indeed almost a dead man for whom every step was torment, he could only work a few hours every other day or so. This forced him to cut out everything but truly vital matters. He did not lose effectiveness thereby; on the contrary, he became, as Churchill called him once, "Lord Heart of the Matter" and accomplished more than anyone else in wartime Washington.

Pruning the Time-Wasters

The three diagnostic questions deal with unproductive and time-consuming activities over which every knowledge worker has some control. Every knowledge worker should ask them. Managers, however, need to be equally concerned with time-loss that results from poor management and deficient organization. Poor management wastes everybody's time—but above all, it wastes the manager's time. Four major time-wasters caused by management and organizational deficiency are discussed below.

1. The first organizational time-wasters result from lack of system or foresight.

The symptom to look for is the recurrent "crisis," the crisis that comes back year after year. A crisis that recurs a second time is a crisis that must not occur again.

The annual inventory crisis belongs here. That with the computer we now can meet it even more "heroically" and at greater expense than we could in the past is hardly a great improvement.

A recurrent crisis should always have been foreseen. It can therefore either be prevented or reduced to a routine that clerks can manage. The definition of a "routine" is that it makes unskilled people without judgment capable of doing what it took near-genius to do before; for a routine puts down in systematic, step-by-step form what a very able man learned in surmounting yesterday's crisis.

The recurrent crisis is not confined to the lower levels of an organization. It afflicts everyone.

For years, a fairly large company ran into one of these crises annually around the first of December. In a highly seasonal business, with the last quarter usually the year's low, fourth-quarter sales and profits were not easily predictable. Every year, however, management made an earnings prediction when it issued its interim report at the end of the second quarter. Three months later, in the fourth quarter, there was tremendous scurrying and companywide emergency action to live up to top management's forecast. For three to five weeks, nobody in the management group got any work done. It took only one stroke of the pen to solve this crisis; instead of predicting a definite year-end figure, top management is now predicting results within a range. That fully satisfies directors, stockholders, and the financial community. And what used to be a crisis a few years ago, now is no longer even noticed in the company—yet fourth-quarter results are quite a bit better than they used to be, since executive time is no longer being wasted on making results fit the forecast.

Prior to Robert McNamara's appointment as secretary of defense in 1961, a similar last-minute crisis shook the entire American defense establishment every spring, toward the end of the fiscal year on June 30. Every manager in the defense establishment, military or civilian, tried desperately in May and June to find expenditures for the money appropriated by Congress for the fiscal year. Otherwise, he was afraid he would have to give back the money. (This last-minute spending spree has also been a chronic disease in Russian planning.) And yet this crisis was totally unnecessary, as

Mr. McNamara immediately saw. The law had always permitted the placing of unspent, but needed, sums into an interim account.

The recurrent crisis is simply a symptom of slovenliness and laziness.

Years ago when I first started out as a consultant, I had to learn how to tell a well-managed industrial plant from a poorly managed one—without any pretense to production knowledge. A well-managed plant, I soon learned, is a quiet place. A factory that is "dramatic," a factory in which the "epic of industry" is unfolded before the visitor's eyes, is poorly managed. A well-managed factory is boring. Nothing exciting happens in it because the crises have been anticipated and have been converted into routine.

Similarly a well-managed organization is a "dull" organization. The "dramatic" things in such an organization are basic decisions that make the future, rather than heroics in mopping up yesterday's mistakes.

2. Time-waste often results from overstaffing.

A workforce may, indeed, be too small for the task. And the work then suffers, if it gets done at all. But this is not the rule. Much more common is the workforce that is too big for effectiveness, the workforce that spends, therefore, an increasing amount of its time "interacting" rather than working.

There is a fairly reliable symptom of overstaffing. If the senior people in the group—and of course the manager in particular—spend more than a small fraction of their time, maybe one-tenth, on "problems of human relations," on feuds and frictions, on jurisdictional disputes and questions of cooperation, and so on, then the workforce is almost certainly too large. People get into each other's way. People have become an impediment to performance, rather than the means thereto. In a lean organization people have room to move without colliding with one another and can do their work without having to explain it all the time.

3. Another common time-waster is malorganization. Its symptom is an excess of meetings.

Meetings are by definition a concession to deficient organization. For one either meets or one works. One cannot do both at the same time. In an ideally designed structure (which in a changing world is of course only a dream), there would be no meetings. Everybody would know what he needs to know to do his job. Everyone would have the resources available to him to do his job. We meet because people holding different jobs have to cooperate to get a specific task done.

But above all, meetings have to be the exception rather than the rule. An organization in which everybody meets all the time is an organization in which no one gets anything done. Wherever a time log shows the fatty degeneration of meeting—whenever, for instance, people in an organization find themselves in meetings a quarter of their time or more—there is time-wasting malorganization.

As a rule, meetings should never be allowed to become the main demand on a knowledge worker's time. Too many meetings always bespeak poor structure of jobs and the wrong organizational components. Too many meetings signify that work that should be in one job or in one component is spread over several jobs or several components. They signify that responsibility is diffused and that information is not addressed to the people who need it.

4. The last major time-waster is malfunction in information.

The administrator of a large hospital was plagued for years by telephone calls from doctors asking him to find a bed for one of their patients who should be hospitalized. The admissions people "knew" that there was no empty bed. Yet the administrator almost invariably found a few. The admissions people simply were not informed immediately when a patient was discharged. The floor nurse knew, of course, and so did the people in the front office who presented the bill to the departing patient. The admissions people, however, got a "bed count" made every morning at 5:00 A.M.—while the great majority of patients were being sent home in midmorning after the doctors had made the rounds. It did not take genius to put this right; all it needed was an extra carbon copy of the chit that goes from the floor nurse to the front office.

Time-wasting management defects such as overstaffing, malorganization, or malfunctioning information can sometimes be remedied fast. At other times, it takes long, patient work to correct them. The results of such work are, however, great—and especially in terms of time gained.

Consolidating "Discretionary Time"

The executive who records and analyzes his time and then attempts to manage it can determine how much he has for his important tasks. How much time is there that is "discretionary," that is, available for the big tasks that will really make a contribution?

It is not going to be a great deal, no matter how ruthlessly the knowledge worker prunes time-wasters.

The higher up a knowledge worker, the larger will be the proportion of time that is not under his control and yet not spent on contribution. The larger the organization, the more time will be needed just to keep the organization together and running, rather than to make it function and produce.

The effective people therefore knows that he has to consolidate his discretionary time. He knows that he needs large chunks of time and that small driblets are no time at all. Even one-quarter of the working day, if consolidated in large time units, is usually enough to get the important things done. But even three-quarters of the working day are useless if it is only available as fifteen minutes here or half an hour there.

The final step in time management is therefore to consolidate the time that record and analysis show as normally available and under the executive's control.

There are a good many ways of doing this. Some people, usually senior managers, work at home one day a week; this is a particularly common method of time consolidation for editors or research scientists.

Others schedule all the operating work—the meetings, reviews, problem sessions, and so on—for two days a week, for example, Monday and Friday, and set aside the mornings of the remaining days for consistent, continuing work on major issues.

But the method by which one consolidates one's discretionary time is far less important than the approach. Most people tackle the job by trying to push the secondary, the less productive matters together, thus clearing, so to speak, a free space between them. This does not lead very far, however. One still gives priority in one's mind and in one's schedule to the less important things, the things that have to be done even though they contribute little. As a result, any new time pressure is likely to be satisfied at the expense of the discretionary time and of the work that should be done in it. Within a few days or weeks, the entire discretionary time will then be gone again, nibbled away by new crises, new immediacies, new trivia.

And all effective people work on their time management perpetually. They not only keep a continuing log and analyze it periodically; they set themselves deadlines for the important activities, based on their judgment of their discretionary time.

One highly effective man I know keeps two such lists—one of the urgent and one of the unpleasant things that have to be done—each with a deadline. When he finds his deadlines slipping, he knows his time is again getting away from him.

Time is the scarcest resource, and unless it is managed, nothing else can be managed. The analysis of one's time, moreover, is the one easily accessible and yet systematic way to analyze one's work and to think through what really matters in it.

"Know thyself," the old prescription for wisdom, is almost impossibly difficult for mortal men. But everyone can follow the injunction "Know thy time" if he or she wants to, and be well on the road toward contribution and effectiveness.

17.

EFFECTIVE DECISIONS

E ffective people do not make a great many decisions. They concentrate on the important ones. They try to think through
what is strategic and generic, rather than "solve problems." They try
to make the few important decisions on the highest level of conceptual understanding. They try to find the constants in a situation.
They are, therefore, not overly impressed by speed in decision-
making. Rather, they consider virtuosity in manipulating a great
many variables a symptom of sloppy thinking. They want to know
what the decision is all about and what the underlying realities are
that it has to satisfy. They want impact rather than technique; they
want to be sound rather than clever.

Effective people know when a decision has to be based on principle and when it should be made on the merits of the case and
pragmatically. They know that the trickiest decision is that between
the right and the wrong compromise and have learned to tell one
from the other. They know that the most time-consuming step in
the process is not making the decision but putting it into effect.
Unless a decision has "degenerated into work," it is not a decision;
it is at best a good intention. This means that, while the effective

decision itself is based on the highest level of conceptual understanding, the action to carry it out should be as close as possible to the working level and as simple as possible.

The least-known of the great American business builders, Theodore Vail, was perhaps the most effective decision-maker in U.S. business history. As president of the Bell Telephone System from just before 1910 till the 1920s, Vail built the organization into the largest private business in the world and into one of the most prosperous growth companies.

Alfred P. Sloan Jr., who in General Motors designed and built the world's largest manufacturing enterprise, took over as head of a big business in 1922, when Vail's career was drawing to its close. He was a very different man, as his was a very different time. And yet the decision for which Sloan is best remembered, the decentralized organizational structure of General Motors, is of the same kind as the major decisions Theodore Vail had made somewhat earlier for the Bell Telephone System.

As Sloan has recounted in his book, *My Years with General Motors,* the company he took over in 1922 was a loose federation of almost independent chieftains. Each of these men ran a unit that a few short years before had still been his own company—and each ran it as if it were still his own company.

Sloan realized that this was not the peculiar and short-term problem of the company just created through merger, but a generic problem of big business.

The Decision Process

The truly important features of the decisions Vail and Sloan made are neither their novelty nor their controversial nature. They are:

1. The clear realization that the problem was generic and could only be solved through a decision that established a rule, a principle

2. The definition of the specifications that the answer to the problem had to satisfy, that is, of the "boundary conditions"

3. The thinking through what is "right," that is, the solution that will fully satisfy the specifications *before* attention is given to the compromises, adaptations, and concessions needed to make the decision acceptable

4. The building into the decision of the action to carry it out

5. The "feedback" that tests the validity and effectiveness of the decision against the actual course of events

These are the *elements* of the effective decision process.

Four Types of Occurrences

1. The first questions the effective decision-maker asks are: Is this a generic situation or an exception? Is this something that underlies a great many occurrences? Or is the occurrence a unique event that needs to be dealt with as such? The generic always has to be answered through a rule, a principle. The exceptional can only be handled as such and as it comes.

Strictly speaking, one might distinguish among four, rather than between two, different types of occurrences.

There is first the truly generic, of which the individual occurrence is only a symptom.

Most of the problems that come up in the course of the executive's work are of this nature. Inventory decisions in a business, for instance, are not "decisions." They are adaptations. The problem is generic. This is even more likely to be true of events within production.

Typically, a product control and engineering group will handle many hundreds of problems in the course of a month. Yet, whenever these are analyzed, the great majority prove to be just symptoms—that is, manifestations of underlying basic conditions. The individual process control engineer or production engineer who works in one part of the plant usually cannot see this. He might have a few problems each month with the couplings in the pipes that carry steam or hot liquids. But only when the total workload of the group over several months is analyzed does the generic problem appear. Then one sees that temperatures or pressures have become too great for the existing equipment and that the couplings, holding different lines together, need to be redesigned for greater loads. Until this is done, process control will spend a tremendous amount of time fixing leaks without ever getting control of the situation.

Then there is the problem that, while a unique event for the individual institution, is actually generic.

The company that receives an offer to merge from another, larger one will never receive such an offer again if it accepts. This is a nonrecurrent situation as far as the individual company, its board of directors, and its management are concerned. But it is, of course, a generic situation that occurs all the time. To think through whether to accept or to reject the offer requires some general rules. For these, however, one has to look to the experience of others.

Next there is the truly exceptional, the truly unique event.

The power failure that plunged into darkness the whole of northeastern North America from the St. Lawrence River to Washington, D.C., in November 1965 was, according to the first explanations, a truly exceptional situation. So was the thalidomide tragedy that led to the birth of so many deformed babies in the early 1960s. The probability of these events, we were told, was one in ten million or one in a hundred million. Such concatenation of malfunctions is as unlikely ever to recur as it is unlikely, for instance, for the chair on which I sit to disintegrate into its constituent atoms.

Truly unique events are rare, however. Whenever one appears, one has to ask, Is this a true exception or only the first manifestation of a new genus?

And this, the early manifestation of a new generic problem, is the fourth and last category of events with which the decision process deals.

We know now, for instance, that both the northeastern power failure and the thalidomide tragedy were only the first occurrences of what, under conditions of modern power technology or of modern pharmacology, are likely to become fairly frequent malfunctions unless generic solutions are found.

All events but the truly unique require a generic solution. They require a rule, a policy, a principle. Once the right principle has been developed, all manifestations of the same generic situation can be handled pragmatically, that is, by adaptation of the rule to the concrete circumstances of the case. Truly unique events, however, must be treated individually. One cannot develop rules for the exceptional.

The effective decision-maker spends time to determine with which of these four situations he is dealing. He knows that he will make the wrong decision if he classifies the situation wrongly.

By far the most common mistake is to treat a generic situation as if it were a series of unique events, that is, to be pragmatic when one lacks the generic understanding and principle. This inevitably leads to frustration and futility.

Specifications of Decision

2. The second major element in the decision process is clear specifications as to what the decision has to accomplish. What are the objectives the decision has to reach? What are the minimum goals it has to attain? What are the conditions it has to satisfy? In science these are known as "boundary conditions." A decision, to be effective, needs to satisfy the boundary conditions. It needs to be adequate to its purpose.

The more concisely and clearly boundary conditions are stated, the greater the likelihood that the decision will indeed be an effective one and will accomplish what it set out to do. Conversely, any serious shortfall in defining these boundary conditions is almost certain to make a decision ineffectual, no matter how brilliant it may seem.

What is the minimum needed to resolve this problem? is the form in which the boundary conditions are usually probed. Can our needs be satisfied? Alfred P. Sloan presumably asked himself when he took command of General Motors in 1922, by removing the autonomy of the division heads. His answer was clearly in the negative. The boundary conditions of his problem demanded strength and responsibility in the chief operating positions. This was needed as much as unity and control at the center. The boundary conditions demanded a solution to a problem of structure, rather than an accommodation among personalities. And this in turn made his solution last.

The effective person knows that a decision that does not satisfy the boundary conditions is ineffectual and inappropriate. It may be worse indeed than a decision that satisfies the wrong boundary conditions. Both will be wrong, of course. But one can salvage the appropriate decision for the incorrect boundary conditions. It is still an effective decision. One cannot get anything but trouble from the decision that is inadequate to its specifications.

In fact, clear thinking about the boundary conditions is needed so that one knows when a decision has to be abandoned.

But clear thinking about the boundary conditions is needed also to identify the most dangerous of all possible decisions: the one that might—just might—work if nothing whatever goes wrong. These decisions always seem to make sense. But when one thinks through the specifications they have to satisfy, one always finds that they are essentially incompatible with each other. That such a decision might succeed is not impossible—it is merely grossly improbable. The trouble with miracles is not, after all, that they happen rarely; it is that one cannot rely on them.

A perfect example was President Kennedy's Bay of Pigs decision in 1961. One specification was clearly Castro's overthrow. But at the same time, there was another specification: not to make it appear that U.S. forces were intervening in one of the American republics. That the second specification was rather absurd, and that no one in the whole world would have believed for one moment that the invasion was a spontaneous uprising of the Cubans, is beside the point. To the American policy-makers at the time, the appearance of nonintervention seemed a legitimate and indeed a necessary condition. But these two specifications would have been compatible with each other only if an immediate islandwide uprising against Castro would have completely paralyzed the Cuban army. And this, while not impossible, was clearly not highly probable in a police state. Either the whole idea should have been dropped or American full-scale support should have been provided to ensure success of the invasion.

It is not disrespect for President Kennedy to say that his mistake was not, as he explained, that he had "listened to the experts." The mistake was failure to think through clearly the boundary conditions that the decision had to satisfy, and refusal to face up to the unpleasant reality that a decision that has to satisfy two different and at bottom incompatible specifications is not a decision but a prayer for a miracle.

Yet, defining the specifications and setting the boundary conditions cannot be done on the "facts" in any decision of importance. It always has to be done on interpretation. It is risk-taking judgment.

Everyone can make the wrong decision—in fact, everyone will sometimes make a wrong decision. But no one needs to make a decision that, on its face, falls short of satisfying the boundary conditions.

What Is Right

3. One has to start out with what is right rather than what is acceptable (let alone who is right) precisely because one always has to compromise in the end. But if one does not know what is right to

satisfy the specifications and boundary conditions, one cannot distinguish between the right compromise and the wrong compromise—and will end up by making the wrong compromise.

I was taught this when I started in 1944 on my first big consulting assignment, a study of the management structure and management policies of the General Motors Corporation. Alfred P. Sloan Jr., who was then chairman and chief executive officer of the company, called me to his office at the start of my study and said, "I shall not tell you what to study, what to write, or what conclusions to come to. This is your task. My only instruction to you is to put down what you think is right as you see it. Don't you worry about our reaction. Don't you worry about whether we will like this or dislike that. And don't you, above all, concern yourself with the compromises that might be needed to make your recommendations acceptable. There is not one executive in this company who does not know how to make every single conceivable compromise without any help from you. But he can't make the *right* compromise unless you first tell him what 'right' is." The executive thinking through a decision might put this in front of himself in neon lights.

For there are two different kinds of compromise. One kind is expressed in the old proverb, Half a loaf is better than no bread. The other kind is expressed in the story of the judgment of Solomon, which was clearly based on the realization that half a baby is worse than no baby at all. In the first instance, the boundary conditions are still being satisfied. The purpose of bread is to provide food, and half a loaf is still food. Half a baby, however, does not satisfy the boundary conditions. For half a baby is not half of a living and growing child. It is a corpse in two pieces.

It is fruitless and a waste of time to worry about what is acceptable and what one had better not say so as not to evoke resistance. The things one worries about never happen. And objections and difficulties no one thought about suddenly turn out to be almost insurmountable obstacles. One gains nothing, in other words, by starting out with the question, What is acceptable? And in the process of answering it, one gives away the important things, as a rule, and loses any chance to come up with an effective, let alone with the right, answer.

Converting into Action

4. Converting the decision into action is the fourth major element in the decision process. While thinking through the boundary conditions is the most difficult step in decision-making, converting the decision into effective action is usually the most time-consuming one. Yet a decision will not become effective unless the action commitments have been built into the decision from the start.

In fact, no decision has been made unless carrying it out in specific steps has become someone's work assignment and responsibility. Until then, there are only good intentions.

This is the trouble with so many policy statements, especially of business: they contain no action commitment. To carry them out is no one's specific work and responsibility. No wonder that the people in the organization tend to view these statements cynically if not as declarations of what top management is really not going to do.

Converting a decision into action requires answering several distinct questions: Who has to know of this decision? What action has to be taken? Who is to take it? And what does the action have to be so that the people who have to do it *can* do it? The first and the last of these are too often overlooked—with dire results.

A story that has become a legend among operations researchers illustrates the importance of the question, Who has to know? A major manufacturer of industrial equipment decided several years ago to discontinue one model. For years it had been standard equipment on a line of machine tools, many of which were still in use. It was decided, therefore, to sell the model to present owners of the old equipment for another three years as a replacement, and then to stop making and selling it. Orders for this particular model had been going down for a good many years. But they shot up as former customers reordered against the day when the model would no longer be available. No one had, however, asked, Who needs to know of this decision? Therefore, nobody informed the clerk in the purchasing department who was in charge of buying the parts from which the model itself was being assembled. His instructions were to buy parts in a given ratio to current sales—and the instructions remained unchanged. When the time came to discontinue further production of the model, the company had in its warehouse enough parts for another eight to ten years of production, parts that had to be written off at a considerable loss.

Feedback

5. Finally, a feedback has to be built into the decision to provide a continual testing, against actual events, of the expectations that underlie the decision.

Decisions are made by human beings who are fallible; at their best their works do not last long. Even the best decision has a high probability of being wrong. Even the most effective one eventually becomes obsolete.

When General Eisenhower was elected president, his predecessor, Harry S. Truman, said, "Poor Ike; when he was a general, he gave an order and it was carried out. Now he is going to sit in that big office and he'll give an order and not a damn thing is going to happen."

The reason why "not a damn thing is going to happen" is, however, not that generals have more authority than presidents. It is that military organizations learned long ago that futility is the lot of most orders and organized the feedback to check on the execution of the order. They learned long ago that to go oneself and look is the only reliable feedback. Reports—all a president is normally able to mobilize—are not much help. All military services have long ago learned that the officer who has given an order goes out and sees for himself whether it has been carried out. At the least he sends one of his own aides—he never relies on what he is told by the subordinate to whom the order was given. Not that he distrusts the subordinate; he has learned from experience to distrust communications.

This is the reason why a battalion commander is expected to go out and taste the food served his men. He could, of course, read the menus and order this or that item to be brought in to him. But no; he is expected to go into the mess hall and take his sample of the food from the same kettle that serves the enlisted men.

With the coming of the computer this will become even more important, for the decision-maker will, in all likelihood, be even further removed from the scene of action. Unless he accepts, as a matter of course, that he had better go out and look at the scene of action, he will be increasingly divorced from reality. All a computer can handle are abstractions. And abstractions can be relied on only if they are constantly checked against the concrete. Otherwise, they are certain to mislead us.

To go and look for oneself is also the best, if not the only, way to test whether the assumptions on which a decision has been made are still valid or whether they are becoming obsolete and need to be thought through again. And one always has to expect the assumptions to become obsolete sooner or later. Reality never stands still very long.

One needs organized information for the feedback. One needs reports and figures. But unless one builds one's feedback around direct exposure to reality—unless one disciplines oneself to go out and look—one condemns oneself to a sterile dogmatism and with it to ineffectiveness.

Opinions Rather Than Facts

A decision is a judgment. It is a choice between alternatives. It is rarely a choice between right and wrong. It is at best a choice between "almost right" and "probably wrong"—but much more often a choice between two courses of action neither of which is provably more nearly right than the other.

Most books on decision-making tell the reader: First find the facts. But executives who make effective decisions know that one does not start with facts. One starts with opinions. These are, of course, nothing but untested hypotheses and, as such, worthless unless tested against reality. To determine what is a fact requires first a decision on the criteria of relevance, especially on the appropriate measurement. This is the hinge of the effective decision, and usually its most controversial aspect.

Finally, the effective decision does not, as so many texts on decision-making proclaim, flow from a consensus on the facts. The understanding that underlies the right decision grows out of the clash and conflict of divergent opinions and out of the serious consideration of competing alternatives.

To get the facts first is impossible. There are no facts unless one has a criterion of relevance. Events by themselves are not facts.

People inevitably start out with an opinion; to ask them to search for the facts first is even undesirable. They will simply do what everyone is far too prone to do anyhow: look for the facts that fit the conclusion they have already reached. And no one has ever failed to find the facts he is looking for. The good statistician knows this and distrusts all figures—he either knows the fellow who found them or he does not know him; in either case he is suspicious.

The only rigorous method, the only one that enables us to test an opinion against reality, is based on the clear recognition that opinions come first—and that this is the way it should be. Then no one can fail to see that we start out with untested hypotheses—in decision-making as in science the only starting point. We know what to do with hypotheses—one does not argue them; one tests them. One finds out which hypotheses are tenable, and therefore worthy of serious consideration, and which are eliminated by the first test against observable experience.

The effective person encourages opinions. But he insists that the people who voice them also think through what it is that the "experiment"—that is, the testing of the opinion against reality— would have to show. The effective person, therefore, asks, What do we have to know to test the validity of this hypothesis? What would the facts have to be to make this opinion tenable? And he makes it a habit—in himself and in the people with whom he works—to think through and spell out what needs to be looked at, studied, and tested. He insists that people who voice an opinion also take responsibility for defining what factual findings can be expected and should be looked for.

Perhaps the crucial question here is, What is the criterion of relevance? This, more often than not, turns on the measurement appropriate to the matter under discussion and to the decision to be reached. Whenever one analyzes the way a truly effective, a truly right, decision has been reached, one finds that a great deal of work and thought went into finding the appropriate measurement.

The effective decision-maker assumes that the traditional measurement is not the right measurement. Otherwise, there would generally be no need for a decision; a simple adjustment would do. The traditional measurement reflects yesterday's decision. That there is need for a new one normally indicates that the measurement is no longer relevant.

The best way to find the appropriate measurement is again to go out and look for the "feedback" discussed earlier—only this is "feedback" before the decision.

In most personnel matters, for instance, events are measured in "averages," such as the average number of lost-time accidents per hundred employees, the average percentage of absenteeism in the whole workforce, or the average illness rate per hundred. But the executive who goes out and looks for himself will soon find that he needs a different measurement. The averages serve the purposes of the insurance company, but they are meaningless, indeed misleading, for personnel management decisions.

The great majority of all accidents occur in one or two places in the plant. The great bulk of absenteeism is in one department. Even illness resulting in absence from work, we now know, is not distributed as an average, but is concentrated in a very small part of the workforce, e.g., young unmarried women. The personnel actions to which dependence on the averages will lead—for instance, the typical plantwide safety campaign—will not produce the desired results, may indeed make things worse.

Finding the appropriate measurement is thus not a mathematical exercise. It is a risk-taking judgment.

Whenever one has to judge, one must have alternatives among which to choose. A judgment in which one can only say yes or no is no judgment at all. Only if there are alternatives can one hope to get insight into what is truly at stake.

Effective people therefore insist on alternatives of measurement— so that they can choose the one appropriate one.

Develop Disagreement

Unless one has considered alternatives, one has a closed mind.

This, above all, explains why effective decision-makers deliberately disregard the second major command of the textbooks on decision-making and create dissension and disagreement, rather than consensus.

Decisions of the kind the executive has to make are not made well by acclamation. They are made well only if based on the clash of conflicting views, the dialogue between different points of view, the choice between different judgments. The first rule in decision-making is that one does not make a decision unless there is disagreement.

Alfred P. Sloan is reported to have said at a meeting of one of his top committees, "Gentlemen, I take it we are all in complete agreement on the decision here." Everyone around the table nodded assent. "Then," continued Mr. Sloan, "I propose we postpone further discussion of this matter until our next meeting to give ourselves time to develop disagreement and perhaps gain some understanding of what the decision is all about."

Sloan was anything but an "intuitive" decision-maker. He always emphasized the need to test opinions against facts and the need to make absolutely sure that one did not start out with the conclusion and then look for the facts that would support it. But he knew that the right decision demands adequate disagreement.

There are three main reasons for the insistence on disagreement.

It is, first, the only safeguard against the decision-maker's becoming the prisoner of the organization. Everybody always wants something from the decision-maker. Everybody is a special pleader, trying—often in perfectly good faith—to obtain the decision he favors. This is true whether the decision-maker is the president of the United States or the most junior engineer working on a design modification.

The only way to break out of the prison of special pleading and preconceived notions is to make sure of argued, documented, thought-through disagreements.

Second, disagreement alone can provide alternatives to a decision. And a decision without an alternative is a desperate gambler's throw, no matter how carefully thought through it might be. There is always a high possibility that the decision will prove wrong—either because it was wrong to begin with or because a change in circumstances makes it wrong. If one has thought through alternatives during the decision-making process, one has something to fall back on, something that has already been thought through, that has been studied, that is understood. Without such an alternative, one is likely to flounder dismally when reality proves a decision to be inoperative.

Above all, disagreement is needed to stimulate the imagination. One does not, to be sure, need imagination to find the right solution to a problem. But then this is of value only in mathematics. In all matters of true uncertainty such as the executive deals with—whether his sphere is political, economic, social, or military—one needs "creative" solutions that create a new situation. And this means that one needs imagination—a new and different way of perceiving and understanding.

Imagination of the first order is, I admit, not in abundant supply. But neither is it as scarce as is commonly believed. Imagination needs to be challenged and stimulated, however, or else it remains latent and unused. Disagreement, especially if forced to be rea-soned, thought through, documented, is the most effective stimulus we know.

The effective decision-maker, therefore, organizes disagreement. This protects him against being taken in by the plausible but false or incomplete. It gives him the alternatives so that he can choose and make a decision, but also so that he is not lost in the fog when his decision proves deficient or wrong in execution. And it forces the imagination—his own and that of his associates. Disagreement converts the plausible into the right and the right into the good decision.

The effective decision-maker does not start out with the assumption that one proposed course of action is right and that all others must be wrong. Nor does he start out with the assumption, I am right and he is wrong. He starts out with the commitment to find out why people disagree.

Effective people know, of course, that there are fools around and that there are mischief-makers. But they do not assume that the man who disagrees with what they themselves see as clear and obvi-ous is, therefore, either a fool or a knave. They know that unless proven otherwise, the dissenter has to be assumed to be reasonably intelligent and reasonably fair-minded. Therefore, it has to be assumed that he has reached his so obviously wrong conclusion because he sees a different reality and is concerned with a different problem. The effective person, therefore, always asks, What does this fellow have to see if his position were, after all, tenable, rational, intelligent? The effective person is concerned first with *understand-ing*. Only then does he even think about who is right and who is wrong.

In a good law office, the beginner, fresh out of law school, is first assigned to drafting the strongest possible case for the other lawyer's client. This is not only the intelligent thing to do before one sits down to work out the case for one's own client. (One has to assume, after all, that the opposition's lawyer knows his business, too.) It is also the right training for a young lawyer. It trains him not to start out with, "I know why my case is right," but with thinking through what it is that the other side must know, see, or take as probable to believe that it has a case at all. It tells him to see the two cases as alternatives. And only then is he likely to understand what his own case is all about. Only then can he make out a strong case in court that his alternative is to be preferred over that of the other side.

Is a Decision Really Necessary?

There is one final question the effective decision-maker asks: Is a decision really necessary? *One* alternative is always the alternative of doing nothing.

Every decision is like surgery. It is an intervention into a system and therefore carries with it the risk of shock. One does not make unnecessary decisions any more than a good surgeon does unnecessary surgery. Individual decision-makers, like individual surgeons, differ in their styles. Some are more radical or more conservative than others. But by and large, they agree on the rules.

One has to make a decision when a condition is likely to degenerate if nothing is done. This also applies with respect to opportunity. If the opportunity is important and is likely to vanish unless one acts with dispatch, one acts—and one makes a radical change.

At the opposite end there are those conditions in respect to which one can, without being unduly optimistic, expect that they will take care of themselves even if nothing is done. If the answer to the question, What will happen if we do nothing? is It will take care of itself, one does not interfere. Nor does one interfere if the condition, while annoying, is of no importance and unlikely to make any difference anyhow.

It is a rare executive who understands this. The controller who in a desperate financial crisis preaches cost reduction is seldom capable of leaving alone minor blemishes, elimination of which will achieve nothing. He may know, for instance, that the significant costs that are out of control are in the sales organization and in physical distribution. And he will work hard and brilliantly at getting them under control. But then he will discredit himself and the whole effort by making a big fuss about the "unnecessary" employment of two or three old employees in an otherwise efficient and well-run plant. And he will dismiss as immoral the argument that eliminating these few semipensioners will not make any difference anyhow. "Other people are making sacrifices," he will argue "Why should the plant people get away with inefficiency?"

When it is all over, the organization will forget fast that he saved the business. They will remember, though, his vendetta against the two or three poor devils in the plant—and rightly so. *De minimis non curat praetor* (The magistrate does not consider trifles) said the Roman law almost two thousand years ago—but many decision-makers still need to learn it.

The great majority of decisions will lie between these extremes. The problem is not going to take care of itself, but it is unlikely to turn into degenerative malignancy either. The opportunity is only for improvement rather than for real change and innovation, but it is still quite considerable. If we do not act, in other words, we will in all probability survive. But if we do act, we may be better off.

In this situation the effective decision-maker compares effort and risk of action to risk of inaction. There is no formula for the right decision here. But the guidelines are so clear that decision in the concrete case is rarely difficult. They are:

- ➤ Act if on balance the benefits greatly outweigh cost and risk.
- ➤ Act or do not act, but do not "hedge" or compromise.

The surgeon who only takes out half the tonsils or half the appendix risks as much infection or shock as if he did the whole job. And he has not cured the condition, has indeed made it worse. He either operates or he doesn't. Similarly, the effective decision-maker either acts or he doesn't act. He does not take half-action. This is the one thing that is always wrong, and the one sure way not to satisfy the minimum specifications, the minimum boundary conditions.

The decision is now ready to be made. The specifications have been thought through, the alternatives explored, the risks and gains weighed. Everything is known. Indeed, it is always reasonably clear by now what course of action must be taken. At this point the decision does indeed almost "make itself."

And it is at this point that most decisions are lost. It becomes suddenly quite obvious that the decision is not going to be pleasant, is not going to be popular, is not going to be easy. It becomes clear that a decision requires courage as much as it requires judgment. There is no inherent reason why medicines should taste horrible—but effective ones usually do. Similarly, there is no inherent reason why decisions should be distasteful—but most effective ones are.

One thing the effective knowledge worker will not do at this point is give in to the cry, "Let's make another study." That is the coward's way—and all the coward achieves is to die a thousand deaths where the brave man dies but one. When confronted with the demand for "another study," the effective executive asks, Is there any reason to believe that additional study will produce anything new? And is there reason to believe that the new is likely to be relevant? And if the answer is no—as it usually is—the effective decision-maker does not permit another study. He does not waste the time of good people to cover up his own indecision.

But at the same time he will not rush into a decision unless he is sure he understands it. Like any reasonably experienced adult, he has learned to pay attention to what Socrates called his "daemon": the inner voice, somewhere in the bowels, that whispers, "Take care." Just because something is difficult, disagreeable, or frightening is no reason for not doing it if it is right. But one holds back—if only for a moment—if one finds oneself uneasy, perturbed, bothered without quite knowing why. "I always stop when things seem out of focus," is the way one of the best decision-makers of my acquaintance puts it.

Nine times out of ten the uneasiness turns out to be over some silly detail. But the tenth time one suddenly realizes that one has overlooked the most important fact in the problem, has made an elementary blunder, or has misjudged altogether. The tenth time one suddenly wakes up at night and realizes—as Sherlock Holmes did in the famous story—that the "most significant thing is that the hound of Baskerville didn't bark."

But the effective decision-maker does not wait long—a few days, at the most a few weeks. If the "daemon" has not spoken by then, he acts with speed and energy whether he likes to or not.

Knowledge workers are not paid for doing things they like to do. They are paid for getting the right things done—most of all in their specific task, the making of effective decisions.

As a result, decision-making can no longer be confined to the very small group at the top. In one way or another almost every knowledge worker in an organization will either have to become a decision-maker himself or will at least have to be able to play an active, an intelligent, and an autonomous part in the decision-making process. What in the past had been a highly specialized function, discharged by a small and usually clearly defined organ—with the rest adapting within a mold of custom and usage—is rapidly becoming a normal if not an everyday task of every single unit in this new social institution, the large-scale knowledge organization. The ability to make effective decisions increasingly determines the ability of every knowledge worker, at least of those in responsible positions, to be effective altogether.

18.

FUNCTIONING
COMMUNICATIONS

We have more attempts at communications today, that is, more attempts to talk to others, and a surfeit of communications media, unimaginable to the men who, around the time of World War I, started to work on the problems of communicating in organizations. Communications in management has become a central concern to students and practitioners in all institutions—business, the military, public administration, hospital, university, and research. In no other area have intelligent men and women worked harder or with greater dedication than psychologists, human relations experts, managers, and management students have worked on improving communications in our major institutions.

Yet communications has proven as elusive as the unicorn. The noise level has gone up so fast that no one can really listen anymore to all that babble about communications. But there is clearly less and less communicating.

We have learned, mostly through doing the wrong things, four fundamentals of communications.

1. Communication is perception.
2. Communication is expectation.
3. Communication makes demands.
4. Communication and information are different and indeed largely opposite—yet interdependent.

Communication Is Perception, Expectation, and Demand

An old riddle posed by the mystics of many religions—the Zen Buddhists, the Sufis of Islam, and the Rabbis of the Talmud—asks, Is there a sound in the forest if a tree crashes down and no one is around to hear it? We now know that the right answer to this is no. There are sound waves. But there is no sound unless someone perceives it. Sound is created by perception. Sound is communication.

This may seem trite; after all, the mystics of old already knew this, for they too always answered that there is no sound unless someone can hear it. Yet the implications of this rather trite statement are great indeed.

First, it means that it is the recipient who communicates. The so-called communicator, the person who emits the communication, does not communicate. He utters. Unless there is someone who hears, there is no communication. There is only noise.

In Plato's *Phaedo,* which, among other things, is also the earliest extant treatise on rhetoric, Socrates points out that one has to talk to people in terms of their own experience, that is, that one has to use carpenters' metaphors when talking to carpenters, and so on. One can communicate only in the recipient's language or in his terms. And the terms have to be experience-based. It, therefore, does very little good to try to explain terms to people. They will not be able to receive them if they are not terms of their own experience. They simply exceed their perception capacity.

In communicating, whatever the medium, the first question has to be, Is this communication within the recipient's range of perception? Can he receive it?

One rarely realizes that there could be other dimensions, and that something that is so obvious to us and so clearly validated by our emotional experience has other dimensions, a "back" and "sides," which are entirely different and which, therefore, lead to entirely different perceptions. The popular story about the blind men and the elephant in which each one, encountering this strange beast, feels one of the elephant's parts, his leg, his trunk, his hide, and reports an entirely different conclusion, and holds to it tenaciously, is simply a metaphor of the human condition. There is no possibility of communication until this is understood and until he who has felt the hide of the elephant goes over to him who has felt the leg and feels the leg himself. There is no possibility of communication, in other words, unless we first know what the recipient, the true communicator, can see and why.

We perceive, as a rule, what we expect to perceive. We see largely what we expect to see, and we hear largely what we expect to hear. That the unexpected may be resented is not the important thing—though most of the work on communications in business and government thinks it is. What is truly important is that the unexpected is usually not received at all. It is not seen or heard, but ignored. Or it is misunderstood, that is, mis-seen or mis-heard as the expected.

The human mind attempts to fit impressions and stimuli into a frame of expectations. It resists vigorously any attempts to make it "change its mind," that is, to perceive what it does not expect to perceive or not to perceive what it expects to perceive. It is, of course, possible to alert the human mind to the fact that what it perceives is contrary to its expectations. But this first requires that we understand what it expects to perceive. It then requires that there be an unmistakable signal—"this is different," that is, a shock that breaks continuity.

Before we can communicate, we must, therefore, know what the recipient expects to see and hear. Only then can we know whether communication can utilize his expectations—and what they are—or whether there is need for the "shock of alienation," for an "awakening" that breaks through the recipient's expectations and forces him to realize that the unexpected is happening.

A phenomenon known to every newspaper editor is the amazingly high readership and retention of the "fillers," the little three- or five-line bits of irrelevant incidental information that are used to "balance" a page. Why should anybody want to read, let alone remember, that it first became fashionable to wear different-colored hose on each leg at the court of some long-forgotten duke? Or, when and where baking powder was first used? Yet there is no doubt that these little tidbits of irrelevancy are read and, above all, that they are remembered far better than almost anything else in the daily paper except the screaming headlines of the catastrophes. The answer is that the fillers make no demands. It is their total irrelevancy that accounts for their being remembered.

Communication, in other words, always makes demands. It always demands that the recipient become somebody, do something, believe something. It always appeals to motivation. If, in other words, communication fits in with the aspirations, the values, the purposes of the recipient, it is powerful. If it goes against his aspirations, his values, his motivations, it is likely not to be received at all or, at best, to be resisted.

Of course, at its most powerful, communication brings about "conversion," that is, a change of personality, of values, beliefs, aspirations. But this is the rare, existential event, and one against which the basic psychological forces of every human being are strongly organized. Even the Lord, the Bible reports, first had to strike Saul blind before he could raise him up as Paul. Communications aiming at conversion demand surrender.

Communication and Information

Where communication is perception, information is logic. As such, information is purely formal and has no meaning. It is impersonal rather than interpersonal. The more it can be freed of the human component, that is, of such things as emotions and values, expectations and perceptions, the more valid and reliable does it become. Indeed, it becomes increasingly informative.

Information presupposes communication. Information is always encoded. To be received, let alone to be used, the code must be known and understood by the recipient. This requires prior agreement, that is, some communication.

Communications, in other words, may not be dependent on information. Indeed, the most perfect communications may be purely "shared experiences," without any logic whatever. Perception has primacy rather than information.

Downward and Upward

What, then, can our knowledge and our experience teach us about communications in organizations, about the reasons for our failures, and about the prerequisites for success in the future?

For centuries we have attempted communication "downward." This, however, cannot work, no matter how hard and how intelligently we try. It cannot work, first, because it focuses on what *we* want to say. It assumes, in other words, that the utterer communicates.

This does not mean that managers should stop working on clarity in what they say or write. Far from it. But it does mean that how we say something comes only after we have learned what to say. And this cannot be found out by "talking to," no matter how well it is being done.

But "listening" does not work either. The Human Relations School of Elton Mayo, forty years ago, recognized the failure of the traditional approach to communications. Its answer was to enjoin listening. Instead of starting out with what "we," that is, the executive, want to "get across," the executive should begin by finding out what subordinates want to know, are interested in, are, in other words, receptive to. To this day, the human relations prescription, though rarely practiced, remains the classic formula.

Of course, listening is a prerequisite to communication. But it is not adequate, and it cannot, by itself, work. Listening assumes that the superior will understand what he is being told. It assumes, in other words, that the subordinates can communicate. It is hard to see, however, why the subordinate should be able to do what his superior cannot do. In fact, there is no reason for assuming he can.

This is not to say that listening is wrong, any more than the futility of downward communications furnishes any argument against attempts to write well, to say things clearly and simply, and to speak the language of those whom one addresses rather than one's own jargon. Indeed, the realization that communications have to be upward—or rather that they have to start with the recipient rather than the emitter, which underlies the concept of listening— is absolutely sound and vital. But listening is only the starting point.

More and better information does not solve the communications problem, does not bridge the communications gap. On the contrary, the more information, the greater is the need for functioning and effective communication. The more information, in other words, the greater is the communications gap likely to be.

Management by Objectives

Can we then say anything constructive about communication? Can we do anything?

Management by objectives is a prerequisite for functioning communication. It requires the subordinate to think through and present to the superior his own conclusions as to what major contribution to the organization—or to the unit within the organization—he should be expected to perform and should be held accountable for.

What the subordinate comes up with is rarely what the superior expects. Indeed, the first aim of the exercise is precisely to bring out the divergence in perception between superior and subordinate. But the perception is focused, and focused on something that is real to both parties. To realize that they see the same reality differently is in itself already communication.

Management by objectives gives to the intended recipient of communication—in this case the subordinate—access to experience that enables him to understand. He is given access to the reality of decision-making, the problems of priorities, the choice between what one likes to do and what the situation demands, and above all, the responsibility for a decision. He may not see the situation the same way the superior does—in fact, he rarely will or even should. But he may gain an understanding of the complexity of the superior's situation and of the fact that the complexity is not of the superior's making, but is inherent in the situation itself.

The examples given in this chapter perhaps illustrate the main conclusion to which our experience with communications—largely an experience of failure—and all the work on learning, memory, perception, and motivation point: communication requires shared experience.

There can be no communication if it is conceived as going from the "I" to the "Thou." Communication works only from one member of "us" to another. Communication in an organization—and this may be the true lesson of our communication failure and the true measure of our communication need—is not a *means* of organization. It is the *mode* of organization.

19.

LEADERSHIP AS WORK

Leadership is all the rage just now. "We'd want you to run a seminar for us on how one acquires charisma," the human-resources VP of a big bank said to me on the telephone—in dead earnest.

Books, articles, and conferences on leadership and on the "qualities" of the leader abound. Every CEO, it seems, has to be made to look like a dashing Confederate cavalry general or a boardroom Elvis Presley.

Leadership does matter, of course. But, alas, it is something different from what is now touted under this label. It has little to do with "leadership qualities" and even less to do with "charisma." It is mundane, unromantic, and boring. Its essence is performance.

In the first place, leadership is not by itself good or desirable. Leadership is a means. Leadership to what end is thus the crucial question.

History knows no more charismatic leaders than this century's triad of Stalin, Hitler, and Mao—the misleaders who inflicted as much evil and suffering on humanity as have ever been recorded.

But effective leadership doesn't depend on charisma. Dwight

Eisenhower, George Marshall, and Harry Truman were singularly effective leaders, yet none possessed any more charisma than a dead mackerel. Nor did Konrad Adenauer, the chancellor who rebuilt West Germany after World War II. No less charismatic personality could be imagined than Abe Lincoln of Illinois, the raw-boned, uncouth backwoodsman of 1860. And there was amazingly little charisma to the bitter, defeated, almost broken Churchill of the interwar years; what mattered was that he turned out in the end to have been right.

Indeed, charisma becomes the undoing of leaders. It makes them inflexible, convinced of their own infallibility, unable to change. This is what happened to Stalin, Hitler, and Mao, and it is a commonplace in the study of ancient history that only Alexander the Great's early death saved him from becoming an ineffectual failure.

Indeed, charisma does not by itself guarantee effectiveness as a leader. John F. Kennedy may have been the most charismatic person ever to occupy the White House. Yet few presidents got as little done.

Nor are there any such things as "leadership qualities" or a "leadership personality." Franklin D. Roosevelt, Winston Churchill, George Marshall, Dwight Eisenhower, Bernard Montgomery, and Douglas MacArthur were all highly effective—and highly visible—leaders during World War II. No two of them shared any "personality traits" or any "qualities."

Work, Responsibility, and Trust Earned

What then is leadership if it is not charisma and not a set of personality traits? The first thing to say about it is that it is work—something stressed again and again by the most charismatic leaders: Julius Caesar, for instance, or General MacArthur and Field Marshal Montgomery, or, to use an example from business, Alfred Sloan, the man who built and led General Motors from 1920 to 1955.

The foundation of effective leadership is thinking through the organization's mission, defining it, and establishing it, clearly and visibly. The leader sets the goals, sets the priorities, and sets and maintains the standards. He makes compromises, of course; indeed, effective leaders are painfully aware that they are not in control of the universe. (Only misleaders—the Stalins, Hitlers, Maos—suffer from that delusion.) But before accepting a compromise, the effective leader has thought through what is right and desirable. The leader's first task is to be the trumpet that sounds a clear sound.

What distinguishes the leader from the misleader are his goals. Whether the compromise he makes with the constraints of reality—which may involve political, economic, financial, or interpersonal problems—are compatible with his mission and goals or lead away from them determines whether he is an effective leader. And whether he holds fast to a few basic standards (exemplifying them in his own conduct), or whether "standards" for him are what he can get away with, determines whether the leader has followers or only hypocritical time-servers.

The second requirement is that the leader see leadership as responsibility rather than as rank and privilege. Effective leaders are rarely "permissive." But when things go wrong—and they always do—they do not blame others. If Winston Churchill is an example of leadership through clearly defining mission and goals, General George Marshall, America's chief of staff in World War II, is an example of leadership through responsibility. Harry Truman's folksy "The buck stops here" is still as good a definition as any.

But precisely because an effective leader knows that he, and no one else, is ultimately responsible, he is not afraid of strength in associates and subordinates. Misleaders are; they always go in for purges. But an effective leader wants strong associates; he encourages them, pushes them, indeed glories in them. Because he holds himself ultimately responsible for the mistakes of his associates and subordinates, he also sees the triumphs of his associates and subordinates as his triumphs, rather than as threats. A leader may be personally vain—as General MacArthur was to an almost pathological degree. Or he may be personally humble—both Lincoln and Truman were so almost to the point of having inferiority complexes. But all three wanted able, independent, self-assured people around them; they encouraged their associates and subordinates, praising and promoting them. So did a very different person: Dwight "Ike" Eisenhower, when supreme commander in Europe.

An effective leader knows, of course, that there is a risk: able people tend to be ambitious. But he realizes that it is a much smaller risk than to be served by mediocrity. He also knows that the gravest indictment of a leader is for the organization to collapse as soon as he leaves or dies, as happened in Russia the moment Stalin died and as happens all too often in companies. An effective leader knows that the ultimate task of leadership is to create human energies and human vision.

The final requirement of effective leadership is to earn trust. Otherwise, there won't be any followers—and the only definition of a leader is someone who has followers. To trust a leader, it is not necessary to like him. Nor is it necessary to agree with him. Trust is the conviction that the leader means what he says. It is a belief in something very old-fashioned, called "integrity." A leader's actions and a leader's professed beliefs must be congruent, or at least compatible. Effective leadership—and again this is very old wisdom—is not based on being clever; it is based primarily on being consistent.

After I had said these things on the telephone to the bank's human-resources VP, there was a long silence. Finally she said, "But that's no different at all from what we have known for years are the requirements for being an effective manager."

Precisely.

20.

PRINCIPLES OF INNOVATION

A ll experienced physicians have seen "miracle cures." Patients suffering from terminal illnesses recover suddenly—sometimes spontaneously, sometimes by going to faith healers, by switching to some absurd diet, or by sleeping during the day and being up and about all night. Only a bigot denies that such cures happen and dismisses them as "unscientific." They are real enough. Yet no physician is going to put miracle cures into a textbook or into a course to be taught to medical students. They cannot be replicated, cannot be taught, cannot be learned. They are also extremely rare; the overwhelming majority of terminal cases do die, after all.

Innovation as a Practice

Similarly, there are innovations that do not proceed from the sources of innovative opportunity, innovations that are not developed in any organized, purposeful, systematic manner. There are innovators who are "kissed by the Muses," and whose innovations are the result of a "flash of genius" rather than of hard, organized, purposeful work. But such innovations cannot be replicated. They cannot be taught and they cannot be learned. There is no known way to teach someone how to be a genius.

But also, contrary to popular belief in the romance of invention and innovation, "flashes of genius" are uncommonly rare. What is worse, I know of not one such "flash of genius" that turned into an innovation. They all remained brilliant ideas.

The purposeful innovation resulting from analysis, system, and hard work is all that can be discussed and presented as the practice of innovation. But this is all that need be presented since it surely covers at least 90 percent of all effective innovations. And the extraordinary performer in innovation, as in every other area, will be effective only if grounded in the discipline and master of it.

What, then, are the principles of innovation, representing the hard core of the discipline? There are a number of "dos"—things that have to be done. There are also a few "don'ts"—things that had better not be done. And then there are what I would call "conditions."

The Dos

1. Purposeful, systematic innovation begins with the analysis of the opportunities. It begins with thinking through what I have called the seven sources of innovative opportunity. In different areas, different sources will have different importance at different times.

> The organization's own unexpected successes and unexpected failures, but also the unexpected successes and unexpected failures of the organization's competitors
> Incongruities, especially incongruities in the process, whether of production or distribution, or incongruities in customer behavior
> Process needs

- Changes in industry and market structures
- Changes in demographics
- Changes in meaning and perception
- New knowledge

All the sources of innovative opportunity should be systematically analyzed and systematically studied. It is not enough to be alerted to them. The search has to be organized, and must be done on a regular, systematic basis.

2. Innovation is both conceptual and perceptual. The second imperative of innovation is therefore to go out to look, to ask, to listen. This cannot be stressed too often. Successful innovators use both the right side and the left side of their brains. They look at figures, and they look at people. They work out analytically what the innovation has to be to satisfy an opportunity. And then they go out and look at the customers, the users, to see what are their expectations, their values, and their needs.

Receptivity can be perceived, as can values. One can perceive that this or that approach will not fit in with the expectations or the habits of the people who have to use it. And then one can ask, What does this innovation have to reflect so that the people who have to use it will *want* to use it, and see in it *their* opportunity? Otherwise, one runs the risk of having the right innovation in the wrong form.

3. An innovation, to be effective, has to be simple and it has to be focused. It should do only one thing; otherwise, it confuses. If it is not simple, it won't work. Everything new runs into trouble; if complicated, it cannot be repaired or fixed. All effective innovations are breathtakingly simple. Indeed, the greatest praise an innovation can receive is for people to say, "This is obvious. Why didn't I think of it?"

Even the innovation that creates new uses and new markets should be directed toward a specific, clear, designed application. It should be focused on a specific need that it satisfies, on a specific end result that it produces.

4. Effective innovations start small. They are not grandiose.

They try to do one specific thing. It may be to enable a moving vehicle to draw electric power while it runs along rails—the innovation that made possible the electric streetcar. Or it may be as elementary as putting the same number of matches into a matchbox (it used to be fifty), which made possible the automatic filling of matchboxes and gave the Swedish originators of the idea a world monopoly on matches for almost half a century. Grandiose ideas, plans that aim at "revolutionizing an industry," are unlikely to work.

Innovations had better be capable of being started small, requiring at first little money, few people, and only a small and limited market. Otherwise, there is not enough time to make the adjustments and changes that are almost always needed for an innovation to succeed. Initially innovations rarely are more than "almost right." The necessary changes can be made only if the scale is small and the requirements for people and money fairly modest.

5. But—and this is the final "do"—a successful innovation aims at leadership. It does not aim necessarily at becoming eventually a "big business"; in fact, no one can foretell whether a given innovation will end up as a big business or a modest achievement. But if an innovation does not aim at leadership from the beginning, it is unlikely to be innovative enough, and therefore unlikely to be capable of establishing itself. Strategies vary greatly, from those that aim at dominance in an industry or a market to those that aim at finding and occupying a small "ecological niche" in a process or market. But all entrepreneurial strategies, that is, all strategies aimed at exploiting an innovation, must achieve leadership within a given environment. Otherwise, they will simply create an opportunity for the competition.

The Don'ts

And now the few important "don'ts."

1. The first is simply not to try to be clever. Innovations have to be handled by ordinary human beings, and if they are to attain any

size and importance at all, by morons or near-morons. Incompetence, after all, is the only thing in abundant and never-failing supply. Anything too clever, whether in design or execution, is almost bound to fail.

2. Don't diversify; don't splinter; don't try to do too many things at once. This is, of course, the corollary to the "do": be focused! Innovations that stray from a core are likely to become diffuse. They remain ideas and do not become innovations. The core does not have to be technology or knowledge. In fact, market knowledge supplies a better core of unity in any enterprise, whether business or public-service institution, than knowledge or technology does. But there has to be a core of unity to innovative efforts or they are likely to fly apart. An innovation needs the concentrated energy of a unified effort behind it. It also requires that the people who put it into effect understand each other, and this, too, requires a unity, a common core. This, too, is imperiled by diversity and splintering.

3. Finally, don't try to innovate for the future. Innovate for the present! An innovation may have long-range impact; it may not reach its full maturity until twenty years later.

But it is not good enough to be able to say, "In twenty-five years there will be so many very old people that they will need this." One has to be able to say, "There are enough old people around today for this to make a difference to them. Of course, time is with us—in twenty-five years there will be many more." But unless there is an immediate application in the present, an innovation is like the drawings in Leonardo da Vinci's notebook—a "brilliant idea." Very few of us have Leonardo's genius and can expect that our notebooks alone will assure immortality.

The first innovator who fully understood this third caveat was probably Thomas Edison. Every other electrical inventor of the time began to work around 1860 or 1865 on what eventually became the light bulb. Edison waited for ten years until the knowledge became available; up to that point, work on the light bulb was "of the future." But when the knowledge became available—when, in other words, a light bulb could become "the present"—Edison organized his tremendous energies and an extraordinarily capable staff and concentrated for a couple of years on that one innovative opportunity.

Innovative opportunities sometimes have long lead times. In pharmaceutical research, ten years of research and development work are by no means uncommon or particularly long. And yet no pharmaceutical company would dream of starting a research project for something that does not, if successful, have immediate application as a drug for health-care needs that already exist.

Three Conditions for a Successful Innovation

Finally, there are three conditions that must be met for an innovation to be successful. All three are obvious but often are disregarded.

1. *Innovation is work.* It requires knowledge. It often requires great ingenuity. There are clearly people who are more talented innovators than the rest of us. Also, innovators rarely work in more than one area. For all his tremendous innovative capacity, Edison worked only in the electrical field. And an innovator in financial areas, Citibank in New York, for instance, is unlikely to embark on innovations in retailing or health care. In innovation as in any other work there is talent, there is ingenuity, and there is predisposition. But when all is said and done, innovation becomes hard, focused, purposeful work making very great demands on diligence, on persistence, and on commitment. If these are lacking, no amount of talent, ingenuity, or knowledge will avail.

2. *To succeed, innovators must build on their strengths.* Successful innovators look at opportunities over a wide range. But then they ask, Which of these opportunities fits *me, fits this company,* puts to work what we (or I) are good at and have shown capacity for in performance? In this respect, of course, innovation is no different from other work. But it may be more important in innovation to build on one's strengths because of the risks of innovation and the result-

ing premium on knowledge and performance capacity. And in innovation, as in any other venture, there must also be a temperamental "fit." Businesses do not do well in something they do not really respect. No pharmaceutical company—run as it has to be by scientifically minded people who see themselves as "serious"—has done well in anything so "frivolous" as lipsticks or perfumes. Innovators similarly need to be temperamentally attuned to the innovative opportunity. It must be important to them and make sense to them. Otherwise they will not be willing to put in the persistent, hard, frustrating work that successful innovation always requires.

3. And finally, *innovation is an effect in economy and society,* a change in the behavior of customers, of teachers, of farmers, of eye surgeons—of people in general. Or it is a change in a process—that is, in how people work and produce something. Innovation therefore always has to be close to the market, focused on the market, indeed market-driven.

The Conservative Innovator

A year or two ago I attended a university symposium on entrepreneurship at which a number of psychologists spoke. Although their papers disagreed on everything else, they all talked of an "entrepreneurial personality," which was characterized by a "propensity for risk-taking."

A well-known and successful innovator and entrepreneur who had built a process-based innovation into a substantial worldwide business in the space of twenty-five years was then asked to comment. He said: "I find myself baffled by your papers. I think I know as many successful innovators and entrepreneurs as anyone, beginning with myself. I have never come across an 'entrepreneurial personality.' The successful ones I know all have, however, one thing—and only one thing—in common: they are *not* 'risk-takers.' They try to define the risks they have to take and to minimize them as much as possible. Otherwise none of us could have succeeded. As for myself, if I had wanted to be a risk-taker, I would have gone into real estate or commodity trading, or I would have become the professional painter my mother wanted me to be."

This jibes with my own experience. I, too, know a good many successful innovators and entrepreneurs. Not one of them has a "propensity for risk-taking."

The popular picture of innovators—half pop-psychology, half Hollywood—makes them look like a cross between Superman and the Knights of the Round Table. Alas, most of them in real life are unromantic figures, and much more likely to spend hours on a cash-flow projection than to dash off looking for "risks."

Of course innovation is risky. But so is stepping into the car to drive to the supermarket for a loaf of bread. All economic activity is by definition "high-risk." And defending yesterday—that is, not innovating—is far more risky than making tomorrow. The innovators I know are successful to the extent to which they define risks and confine them. They are successful to the extent to which they systematically analyze the sources of innovative opportunity, then pinpoint the opportunity and exploit it—whether an opportunity of small and clearly definable risk, such as exploiting the unexpected or a process need, or an opportunity of much greater but still definable risk, as in knowledge-based innovation.

Successful innovators are conservative. They have to be. They are not "risk-focused"; they are "opportunity-focused."

21.

THE SECOND HALF
OF YOUR LIFE

For the first time in human history, individuals can expect to outlive organizations. This creates a totally new challenge: *What to do with the second half of one's life?*

One can no longer expect that the organization for which one works at age thirty will still be around when one reaches age sixty. But also, forty or fifty years in the same kind of work is much too long for most people. They deteriorate, get bored, lose all joy in their work, "retire on the job," and become a burden to themselves and to everyone around them.

This is not necessarily true of the very top achievers such as very great artists. Claude Monet (1840–1926), the greatest Impressionist painter, was still painting masterpieces in his eighties, and working twelve hours a day, even though he had lost almost all his eyesight. Pablo Picasso (1881–1973), perhaps the greatest Postimpressionist painter, similarly painted till he died in his nineties—and in his seventies invented a new style. The greatest musical instrumentalist of this century, the Spanish cellist Pablo Casals (1876–1973), planned to perform a new piece of music and practiced it on the very day on which he died at age ninety-seven. But these are the rarest of exceptions even among very great achievers. Neither Max Planck (1858–1947) nor Albert Einstein (1879–1955), the two giants of modern physics, did important scientific work after their forties. Planck had *two* more careers. After 1918—aged sixty—he reorganized German science. After being forced into retirement by the Nazis in 1933, he, in 1945, almost ninety, started once more to rebuild German science after Hitler's fall. But Einstein retired in his forties to become a "famous man."

There is a great deal of talk today about the "midlife crisis" of the executive. It is mostly boredom. At age forty-five most executives have reached the peak of their business career and know it. After twenty years of doing very much the same kind of work, they are good at their jobs. But few are learning anything anymore, few are contributing anything anymore, and few expect the job again to become a challenge and a satisfaction.

Manual workers who have been working for forty years—in the steel mill for instance, or in the cab of a locomotive—are physically and mentally tired long before they reach the end of their normal life expectancy, that is, well before they reach even traditional retirement age. They are "finished." If they survive—and their life expectancy too has gone up to an average of seventy-five years or so—they are quite happy spending ten or fifteen years doing nothing, playing golf, going fishing, engaging in some minor hobby, and so on. But knowledge workers are not "finished." They are perfectly capable of functioning despite all kinds of minor complaints. And yet the original work that was so challenging when the knowledge worker was thirty has become a deadly bore when the knowledge worker is fifty—and still he or she is likely to face another fifteen if not another twenty years of work.

To manage oneself, therefore, will increasingly require preparing oneself for the second half of one's life.

Three Answers for the Second Half of Life

There are three answers.

The first is actually to start a second and different career (as Max Planck did). Often this means only moving from one kind of organization to another.

Typical are the middle-level American business executives who in substantial numbers move to a hospital, a university, or some other nonprofit organization, around age forty-five or forty-eight, when the children are grown and the retirement pension is vested. In many cases they stay in the same kind of work. The divisional controller in the big corporation becomes, for instance, controller in a medium-sized hospital. But there are also a growing number of people who actually move into a different line of work.

In the United States there are a fairly substantial number of middle-aged women who have worked for twenty years, in business or in local government, have risen to a junior management position, and now, at age forty-five and with the children grown, enter law school. Three or four years later they then establish themselves as small-time lawyers in their local communities.

We will see many more such second-career people who have achieved fair success in their first job. Those people have substantial skills, for example, the divisional controller who moves into the local community hospital. They know how to work. They need a community—and the house is empty with the children gone. They need the income, too. But above all, they need the challenge.

The second answer to the question of what to do with the second half of one's life is to develop a *parallel* career.

A large and rapidly growing number of people—especially people who are very successful in their first careers—stay in the work they have been doing for twenty or twenty-five years. Many keep on working forty or fifty hours a week in their main and paid job. Some move from being busy full-time to being part-time employees or become consultants. But then they create for themselves a parallel job—usually in a nonprofit organization—and one that often takes another ten hours of work a week.

And then, finally—the third answer—there are the "social entrepreneurs." These are usually people who have been very successful in their first profession, as businesspeople, as physicians, as consultants, as university professors. They love their work, but it no longer challenges them. In many cases they keep on doing what they have been doing all along, though they spend less and less of their time on it. But they *start* another, and usually a nonprofit, activity.

People who manage the "second half" may always be a minority. The majority may keep doing what they are doing now, that is, retire on the job, continue being bored, keeping on with their routine, and counting the years until retirement. But it is this minority, the people who see the long working-life expectancy as an opportunity both for themselves and for society, who will increasingly become the leaders and the models. They, increasingly, will be the "success stories."

There is one requirement for managing the second half of one's life: to begin creating it long before one enters it.

When it first became clear thirty years ago that working-life expectancies were lengthening very fast, many observers (including myself) believed that retired people would increasingly become volunteers for American nonprofit institutions. This has not happened. If one does not begin to volunteer before one is forty or so, one will not volunteer when past sixty.

Similarly, all the social entrepreneurs I know began to work in their chosen second enterprise long before they reached their peak in their original business. One highly successful lawyer, for example, began to do volunteer legal work for the schools in his state when he was around thirty-five. He got himself elected to a school board at age forty. When he reached fifty, and had amassed a substantial fortune, he then started his own enterprise to build and run model schools. He is, however, still working nearly full-time as the lead counsel in the very big company that, as a very young lawyer, he had helped found.

There is another reason that managing yourself will increasingly mean that the knowledge worker develops a second *major interest,* and develops it early.

No one can expect to live very long without experiencing a serious setback in one's life or in one's work. There is the competent engineer who at age forty-two is being passed over for promotion in the company. There is the competent college professor who at age forty-two realizes that she will stay forever in the small college in which she got her first appointment and will never get the professorship at the big university—even though she may be fully qualified for it. There are tragedies in one's personal family life—the breakup of one's marriage, the loss of a child.

And then a second major interest—and not just another hobby—may make all the difference. The competent engineer passed over for promotion now knows that he has not been very successful in his job. But in his outside activity—for example, as treasurer in his local church—he has achieved success and continues to have success. One's own family may break up, but in that outside activity there is still a community.

This will be increasingly important in a society in which *success* has become important.

Historically, there was no such thing. The overwhelming majority of people did not expect anything but to stay in their "proper station," as an old English prayer has it. The only mobility there was downward mobility. Success was practically unknown.

In a knowledge society we expect everyone to be a "success." But this is clearly an impossibility. For a great many people there is, at best, absence of failure. For where there is success, there has to be failure. And then it is vitally important for the individual—but equally for the individual's family—that there be an area in which the individual contributes, makes a difference, and is *somebody.* That means having a second area, whether a second career, a parallel career, a social venture, a serious outside interest, anything offering an opportunity for being a leader, for being respected, for being a success.

Revolution for the Individuals

The changes and challenges of managing oneself may seem obvious, if not elementary. And the answers may seem to be self-evident to the point of appearing naive.

Managing oneself, however, is a *revolution* in human affairs. It requires new and unprecedented things from the individual, and especially from the knowledge worker. For in effect, it demands that each knowledge worker *think* and behave as a *chief executive officer.* It also requires an almost 180-degree change in the knowledge worker's thoughts and actions from what most of us—even of the younger generation—still take for granted as the way to think and the way to act. Knowledge workers, after all, first came into being in any substantial numbers a generation ago. (I coined the term "knowledge worker" years ago.)

But also the shift from manual workers who do as they are being told—either by the task or by the boss—to knowledge workers who have to manage themselves profoundly challenges social structure. For every existing society, even the most "individualist" one, takes two things for granted, if only subconsciously: organizations outlive workers, and most people stay put. Managing oneself is based two very opposite *realities:* workers are likely to outlive organizations, and the knowledge worker has mobility.

In the United States, *mobility* is accepted. But even in the United States, workers outliving organizations—and with it the need to be prepared for a *different second half of one's life*—is a revolution for which practically no one is prepared. Nor is any existing institution, for example, the present retirement system. In the rest of the developed world, however, *immobility* is expected and accepted. It is "stability."

In Germany, for instance, mobility—until very recently—came

to an end with the individual's reaching age ten or, at the latest, age sixteen. If a child did not enter *Gymnasium* at age ten, he or she had lost any chance ever to go to the university. And the apprenticeship that the great majority who did not go to the *Gymnasium* entered at age fifteen or sixteen as a mechanic, a bank clerk, a cook—irrevocably and irreversibly—decided what work the person was going to do the rest of his or her life. Moving from the occupation of one's apprenticeship into another occupation was simply not done even when not actually forbidden.

Transformation of Every Society

The developed society that faces the greatest challenge and will have to make the most difficult changes is the society that has been most successful in the last fifty years: Japan. Japan's success—and there is no precedent for it in history—very largely rested on *organized immobility*—the immobility of "lifetime employment." In lifetime employment it is the organization that manages the individual. And it does so, of course, on the assumption that the individual has no choice. The individual is being managed.

I very much hope that Japan will find a solution that *preserves* the social stability, the community—and the social harmony—that lifetime employment provided, and yet creates the mobility that knowledge work and knowledge workers must have. Far more is at stake than Japan's own society and civic harmony. A Japanese solution would provide a model—for in every country a functioning society does require cohesion. Still, a successful Japan will be a very different Japan.

But so will be every other developed country. The emergence of the knowledge worker who both *can* and *must* manage himself or herself is transforming every society.

22.

THE EDUCATED PERSON

K nowledge is not impersonal, like money. Knowledge does not reside in a book, a databank, a software program; they contain only information. Knowledge is always embodied in a person; carried by a person; created, augmented, or improved by a person; applied by a person; taught and passed on by a person; used or misused by a person. The shift to the knowledge society therefore puts the person in the center. In so doing, it raises new challenges, new issues, new and quite unprecedented questions about the knowledge society's representative, the educated person.

In all earlier societies, the educated person was an ornament. He or she embodied *Kultur*—the German term that with its mixture of awe and derision is untranslatable into English (even "highbrow" does not come close). But in the knowledge society, the educated person is society's emblem; society's symbol; society's standard-bearer. The educated person is the social "archetype"—to use the sociologist's term. He or she defines society's performance capacity. But he or she also embodies society's values, beliefs, commitments. If the feudal knight was the clearest embodiment of society in the early Middle Ages, and the "bourgeois" in the Age of Capitalism, the educated person will represent society in the postcapitalist world in which knowledge has become the central resource.

This must change the very meaning of "educated person." It must change the very meaning of what it means to be educated. It will thus predictably make the definition of an "educated person" a crucial issue. With knowledge becoming the key resource, the educated person faces new demands, new challenges, new responsibilities. *The educated person now matters.*

For the last ten or fifteen years a vigorous—often shrill—debate has been raging in American academia over the educated person. Should there be one? Could there be one? And what should be considered "education" anyway?

A motley crew of post-Marxists, radical feminists, and other "antis" argues that there can be no such thing as an educated person—the position of those new nihilists, the "deconstructionists." Others in this group assert that there can be only educated persons with each sex, each ethnic group, each race, each "minority" requiring its own separate culture and a separate—indeed, an isolationist—educated person. Since these people are mainly concerned with the "humanities," there are few echoes as yet of Hitler's "Aryan physics," Stalin's "Marxist genetics," or Mao's "Communist psychology." But the arguments of these antitraditionalists recall those of the totalitarians. And their target is the same: the universalism that is at the very core of the concept of an educated person, whatever it may be called ("educated person" in the West, or *"bunjin"* in China and Japan).

The opposing camp—we might call them the "humanists"— also scorns the present system. But it does so because it fails to produce a universally educated person. The humanist critics demand a return to the nineteenth century, to the "liberal arts," the "classics," the German *Gebildete Mensch.* They do not, so far, repeat the assertion made by Robert Hutchins and Mortimer Adler fifty years ago at the University of Chicago that knowledge in its entirety consists of a hundred "great books." But they are in direct line of descent from the Hutchins-Adler "return to premodernity."

Both sides, alas, are wrong.

At the Core of the Knowledge Society

The knowledge society *must* have at its core the concept of the educated person. It will have to be a universal concept, precisely because the knowledge society is a society of knowledges and because it is global—in its money, its economics, its careers, its technology, its central issues, and above all, in its information. Postcapitalist society requires a unifying force. It requires a leadership group, which can focus local, particular, separate traditions onto a common and shared commitment to values, a common concept of excellence, and on mutual respect.

The postcapitalist society—the knowledge society—thus needs exactly the opposite of what deconstructionists, radical feminists, or anti-Westerners propose. It needs the very thing they totally reject: a universally educated person.

Yet the knowledge society needs a kind of educated person different from the ideal for which the humanists are fighting. They rightly stress the folly of their opponents' demand to repudiate the Great Tradition and the wisdom, beauty, and knowledge that are the heritage of mankind. But a bridge to the past is not enough—and that is all the humanists offer. The educated person needs to be able to bring his or her knowledge to bear on the present, not to mention to have a role in molding the future. There is no provision for such ability in the proposals of the humanists, indeed, no concern for it. But without it, the Great Tradition remains dusty antiquarianism.

In his 1943 novel *Das Glasperlenspiel (The Glass Bead Game)*, Hermann Hesse anticipated the sort of world the humanists want—and its failure. The book depicts a brotherhood of intellectuals, artists, and humanists who live a life of splendid isolation, dedicated to the Great Tradition, its wisdom and its beauty. But the hero, the most accomplished Master of the Brotherhood, decides in the end to return to the polluted, vulgar, turbulent, strife-torn, money-grubbing reality—for his values are only fool's gold unless they have relevance to the world.

What Hesse foresaw more than fifty years ago is now in fact happening. "Liberal education" and *"Allgemeine Bildung"* ("general education") are in crisis today because they have become a *Glasperlenspiel,* which the brightest desert for crass, vulgar, money-grubbing reality. The ablest students appreciate the liberal arts. They enjoy them fully as much as did their great-grandparents, who graduated before World War I. For that earlier generation, liberal arts and *Allgemeine Bildung* remained meaningful throughout their lives, and defined their identity. They still remained meaningful for many members of my generation, which graduated before World War II—even though we immediately forgot our Latin and Greek. But all over the world today's students, a few years after they have graduated, complain that "what I have learned so eagerly has no meaning; it has no relevance to anything I am interested in or want to become." They still want a liberal arts curriculum for their own children—Princeton or Carleton; Oxbridge; Tokyo University; the *lycée;* the *Gymnasium*—though mainly for social status and access to good jobs. But in their own lives they repudiate such values. They repudiate the educated person of the humanists. Their liberal education, in other words, does not enable them to understand reality, let alone to master it.

Both sides in the present debate are largely irrelevant. Postcapitalist society needs the educated person even more than any earlier society did, and access to the great heritage of the past will have to be an essential element. But this heritage will embrace a good deal more than the civilization that is still mainly Western, the Judeo-Christian tradition, for which the humanists are fighting. The educated person we need will have to be able to appreciate other cultures and traditions: for example, the great heritage of Chinese,

Japanese, and Korean paintings and ceramics; the philosophers and religions of the Orient; and Islam, both as a religion and as a culture. The educated person also will have to be far less exclusively "bookish" than the product of the liberal education of the humanists. He or she will need trained perception fully as much as analysis.

The Western tradition will, however, still have to be at the core, if only to enable the educated person to come to grips with the present, let alone the future. The future may be "post-Western"; it may be "anti-Western." It cannot be "non-Western." Its material civilization and its knowledges all rest on Western foundations: Western science; tools and technology; production; economics; Western-style finance and banking. None of these can work unless grounded in an understanding and acceptance of Western ideas and of the entire Western tradition.

The most profoundly "anti-Western" movement today is not fundamentalist Islam. It is the revolt of the "Shining Path" in Peru—the desperate attempt of the descendants of the Incas to undo the Spanish Conquest, to go back to the Indians' ancient tongues of Quechua and Aymara, and to drive the hated Europeans and their culture back into the ocean. But this anti-Western rebellion finances itself by growing coca for the drug addicts of New York and Los Angeles. Its favorite weapon is not the Incas' slingshot; it is the car bomb.

Tomorrow's educated person will have to be prepared for life in a global world. It will be a "Westernized" world, but also increasingly a tribalized world. He or she must become a "citizen of the world"—in vision, horizon, information. But he or she will also have to draw nourishment from their local roots and, in turn, enrich and nourish their own local culture.

Knowledge Society and Society of Organizations

Postcapitalist society is both a knowledge society and a society of organizations, each dependent on the other and yet each very different in its concepts, views, and values. Most, if not all, educated persons will practice their knowledge as members of an organization. The educated person will therefore have to be prepared to live and work simultaneously in two cultures—that of the "intellectual," who focuses on words and ideas, and that of the "manager," who focuses on people and work.

Intellectuals see the organization as a tool; it enables them to practice their *techné*, their specialized knowledge. Managers see knowledge as a means to the end of organizational performances. Both are right. They are opposites; but they relate to each other as poles rather than as contradictions. They surely need each other: the research scientist needs the research manager just as much as the research manager needs the research scientist. If one overbalances the other, there is only nonperformance and all-around frustration. The intellectual's world, unless counterbalanced by the manager, becomes one in which everybody "does his own thing" but nobody achieves anything. The manager's world, unless counterbalanced by the intellectual, becomes the stultifying bureaucracy of the "organization man." But if the two balance each other, there can be creativity and order, fulfillment and mission.

A good many people in the postcapitalist society will actually live and work in these two cultures at the same time. And many more should be exposed to working experience in both cultures, by rotation early in their careers—from a specialist's job to a managerial one, for instance, rotating the young computer technician into project manager and team leader, or by asking the young college professor to work part-time for two years in university administration. And again, working as "unpaid staff" in an agency of the social sector will give the individual the perspective and the balance to respect both worlds, that of the intellectual and that of the manager.

All educated persons in the postcapitalist society will have to be prepared to *understand* both cultures.

Technés and the Educated Person

For the educated person in the nineteenth century, *technés* were not knowledge. They were already taught in the university and had become "disciplines." Their practitioners were "professionals," rather than "tradesmen" or "artisans." But they were not part of the liberal arts or the *Allgemeine Bildung,* and thus not part of knowledge.

University degrees in *technés* go back a long way: in Europe, with both the law degree and the medical degree, as far as the thirteenth century. And on the Continent and in America—though not in England—the new engineering degree (first awarded in Napoleon's France a year or two before 1800) soon became socially accepted. Most people who were considered "educated" made their living practicing a *techné*—whether as lawyers, physicians, engineers, geologists, or, increasingly, in business (only in England was there esteem for the "gentleman" without occupation). But their job or their profession was seen as a "living," not a "life."

Outside their offices, the *techné* practitioners did not talk about their work or even about their disciplines. That was "shop talk"; the Germans sneered at it as *Fachsimpeln.* It was even more derided in France: anyone who indulged in shop talk there was considered a boor and a bore, and promptly taken off the invitation lists of polite society.

But now that the *technés* have become knowledges in the plural, they have to be integrated into knowledge. The *technés* have to become part of what it means to be an educated person. The fact that the liberal arts curriculum they enjoyed so much in their college years refuses to attempt this is the reason why today's students repudiate it a few years later. They feel let down, even betrayed. They have good reason to feel that way. Liberal arts and *Allgemeine Bildung* that do not integrate the knowledges into a "universe of knowledge" are neither "liberal" nor "*Bildung.*" They fall down on their first task: to create mutual understanding, that "universe of discourse" without which there can be no civilization. Instead of uniting, such disciplines only fragment.

We neither need nor will get "polymaths," who are at home in many knowledges; in fact, we will probably become even more specialized. But what we do need—and what will define the educated person in the knowledge society—is the ability to *understand* the various knowledges. What is each one about? What is it trying to do? What are its central concerns and theories? What major new insights has it produced? What are its important areas of ignorance, its problems, its challenges?

To Make Knowledges the Path to Knowledge

Without such understanding, the knowledges themselves will become sterile, will indeed cease to be "knowledges." They will become intellectually arrogant and unproductive. For the major new insights in every one of the specialized knowledges arise out of another, separate specialty, out of another one of the knowledges.

Both economics and meteorology are being transformed at present by the new mathematics of chaos theory. Geology is being profoundly changed by the physics of matter, archaeology by the genetics of DNA typing; history by psychological, statistical, and technological analyses and techniques. An American, James M. Buchanan (b. 1919), received the 1986 Nobel Prize in Economics for applying recent economic theory to the political process and thereby standing on their heads the assumptions and theories on which political scientists had based their work for over a century.

The specialists have to take responsibility for making both themselves and their specialty understood. The media, whether magazines, movies, or television, have a crucial role to play. But they cannot do the job by themselves. Nor can any other kind of popularization. Specialties must be understood for what they are:

serious, rigorous, demanding disciplines. This requires that the leaders in each of the knowledges, beginning with the leading scholars in each field, must take on the hard work of defining what it is they do.

There is no "queen of the knowledges" in the knowledge society. All knowledges are equally valuable; all knowledges, in the words of the great medieval philosopher Saint Bonaventura, lead equally to the truth. But to make them paths to truth, paths to knowledge, has to be the responsibility of the men and women who own these knowledges. Collectively, they hold knowledge in trust.

Capitalism had been dominant for over a century when Karl Marx in the first volume of *Das Kapital* identified it (in 1867) as a distinct social order. The term "capitalism" was not coined until thirty years later, well after Marx's death. It would therefore not only be presumptuous in the extreme to attempt to write *The Knowledge* today; it would be ludicrously premature. All that can be attempted is to describe society and polity as we begin the transition from the age of capitalism (also, of course, the age of socialism).

But we can hope that a hundred years hence a book of this kind, if not one entitled *The Knowledge,* can be written. That would mean that we have successfully weathered the transition upon which we have only just embarked. It would be as foolish to predict the knowledge society as it would have been foolish to predict in 1776—the year of the American Revolution, of Adam Smith's *Wealth of Nations,* and of James Watt's steam engine—the society of which Marx wrote a hundred years later. And it was as foolish of Marx to try to predict in mid-Victorian capitalism—and with "scientific infallibility"—the society in which we live now.

But one thing we can predict: the greatest change will be the change in knowledge—in its form and content; in its meaning; in its responsibility; and in what it means to be an educated person.

III.

SOCIETY

23.

A CENTURY OF SOCIAL TRANSFORMATION—EMERGENCE OF KNOWLEDGE SOCIETY

No century in human history has experienced so many social transformations and such radical ones as the twentieth century. They, I submit, will turn out to be the most significant events of this century, and its lasting legacy. In the developed free-market countries—only one-fifth of the earth's population, but the model for the rest—work and workforce, society and polity, are all, in the last decade of this century, *qualitatively* and *quantitatively* different both from those of the first years of this century and from anything ever experienced before in human history: different in their configuration, in their processes, in their problems, and in their structures.

Far smaller and far slower social changes in earlier periods triggered violent intellectual and spiritual crises, rebellions, and civil wars. The extreme social transformations of this century have hardly caused any stir. They proceeded with a minimum of friction, with a minimum of upheavals, and indeed with altogether a minimum of attention from scholars, politicians, the press, and the public.

To be sure, this century of ours may well have been the cruelest and most violent in human history, with its world wars and civil wars, its mass tortures, ethnic cleansings, and genocides. But all these killings, all these horrors inflicted on the human race by this century's *Weltbeglücker*—those who establish paradise on earth by killing off nonconformists, dissidents, resisters, and innocent bystanders, whether Jews, the bourgeoisie, kulaks, or intellectuals—hindsight clearly shows, were just that: senseless killings, senseless horrors Hitler, Stalin, and Mao, the three evil geniuses of this century, destroyed. But they created nothing.

Indeed, if this century proves anything, it is the futility of politics. Even the most dogmatic believer in historical determinism would have a hard time explaining the social transformations of this century as caused by the headline-making political events, or explaining the headline-making political events as caused by the social transformations. But it is the social transformations, running like ocean currents deep below the hurricane-tormented surface of the sea, that have had the lasting, indeed the permanent, effect. They—rather than all the violence of the political surface—have transformed the society and the economy, the community, the polity we live in.

Farmers and Domestic Servants

Before World War I, the largest single group in every country were farmers.

Eighty years ago, on the eve of that war, it was considered axiomatic that developed countries—North America being the only exception—would increasingly become unable to feed themselves and would increasingly have to rely on food imports from nonindustrial, nondeveloped areas.

Today, only Japan, among major, developed, free-market countries, is a heavy importer of food. (Unnecessarily so—its weakness as a food producer is largely the result of an obsolete rice-subsidy policy that prevents the country from developing a modern, productive agriculture.) All other developed free-market countries have become surplus food producers despite burgeoning urban populations. In all these countries food production is today many times what it was eighty years ago—in the United States, eight to ten times as much.

But in all developed free-market countries—including Japan—farmers today are, at most, 5 percent of population and workforce, that is, one-tenth of what they were eighty years ago.

The second-largest group in population and workforce in every developed country around 1900 were live-in servants. They were considered as much a "law of nature" as farmers were. The British census of 1910 defined "lower middle class" as a household employing fewer than three servants. And while farmers as a proportion of population and workforce had been steadily shrinking throughout the nineteenth century, the numbers of domestic servants, both absolutely and as a percentage, were steadily growing right up to World War I. Eighty years later, live-in domestic servants in developed countries have become practically extinct. Few people born since World War II, that is, few people under fifty, have even seen any except on the stage or in old films.

Farmers and domestic servants were not only the *largest* social groups, they were the oldest social groups, too. Together they were, through the ages, the foundation of economy and society, the foundation altogether of "civilization."

The Rise and Fall of the Blue-collar Worker

One reason, indeed the main reason, why the transformation caused so little stir was that by 1900 a new class, the blue-collar worker in manufacturing industry (Marx's "proletarian"), had become socially dominant. Early-twentieth-century society was obsessed with blue-collar workers, fixated on them, bewitched by them.

The blue-collar worker became the "social question" of 1900 because he was the first "lower class" in history that could be organized and stay organized.

No class in history has ever risen faster than the blue-collar worker. And no class in history has ever fallen faster.

In 1883, the year of Marx's death, "proletarians" were still a minority of industrial workers. The majority were then skilled workers employed in small craft shops each containing twenty or thirty workers at most.

By 1900, the term "industrial worker" had become synonymous with "machine operator" in a factory employing hundreds, if not thousands, of people. These factory workers were indeed Marx's proletarians, without social position, without political power, without economic or purchasing power.

The workers of 1900—and even of 1913—had no pension; no paid vacation; no overtime pay; no extra pay for Sunday or night work; no health insurance (except in Germany); no unemployment compensation; no job security whatever. One of the earliest laws to limit working hours for adult males—enacted in Austria in 1884—set the working day at *eleven* hours, six days a week. Industrial workers, in 1913, everywhere worked a minimum of three thousand hours a year. Their unions were still officially proscribed or, at best, barely tolerated. But the workers had shown their capacity to be organized. They had shown their capacity to act as a "class."

In the 1950s industrial blue-collar workers had become the largest single group in every developed country, including the Communist ones, though they were an actual majority only during wartime. They had become eminently respectable. In all developed free-market countries they had economically become "middle class." They had extensive job security; pensions; long, paid vacations; comprehensive unemployment insurance or "lifetime employment." Above all, they had achieved political power. It was not only in Britain that the labor unions were considered to be the "real government," with greater power than the prime minister and Parliament.

In 1990, however, both the blue-collar worker and his union were in total and irreversible retreat. They had become marginal in numbers. Whereas blue-collar workers who made or moved things had accounted for two-fifths of the American workforce in the 1950s, they accounted for less than one-fifth of the workforce in the early 1990s—that is, for no more than they had accounted for in 1900, when their meteoric rise had begun. In the other developed free-market countries the decline was slower at first; but after 1980 it began to accelerate everywhere. By the year 2000 or 2010, in every developed free-market country, blue-collar industrial workers will account for no more than one-tenth or, at most, one-eighth of the workforce. Union power has been going down equally fast. Where in the 1950s and 1960s the National Union of Mineworkers in the United Kingdom broke prime ministers as if they were matchsticks, Margaret Thatcher in the 1980s won election after election by being openly contemptuous of organized labor and by whittling down its political power and its privileges. The blue-collar worker in manufacturing industry and his union are going the way of the farmer.

His place is already being taken by a "technologist," that is, by people who work both with their hands and their theoretical knowledge. (Examples are computer technicians or paramedical technicians such as X-ray technicians, physical therapists, medical-lab technicians, pulmonary technicians, and so on, who have been the fastest-growing group in the United States workforce since 1980.)

And instead of a "class," that is, a coherent, recognizable, defined, and self-conscious group, the blue-collar worker in manufacturing industry may soon be just another "pressure group."

In contrast with Marxist and syndicalist predictions, the rise of the industrial worker did not destabilize society. On the contrary, it emerged as the century's most *stabilizing social development*. It explains why the disappearance of farmer and domestic servant produced no social crises.

For farmer and domestic servant, industrial work was an opportunity. It was in fact the first opportunity in social history to better oneself substantially without having to emigrate. In the developed, free-market countries, every generation in the last 100 or 150 years could expect to do substantially better than the generation preceding it. The main reason was that farmers and domestic servants could and did become industrial workers.

Because industrial workers were concentrated in groups, that is, because they worked in a large factory rather than in a small shop or in their homes, there could be a systematic focus on their *productivity*. Beginning in 1881—two years before Marx's death—the systematic study of work, of both tasks and tools, has raised the productivity of manual work (the making and moving of things) by 3 to 4 percent, compounded each year, for a total fiftyfold increase in output per worker over a hundred years. On this rest all the economic and social gains of the past century. And contrary to what "everybody knew" in the nineteenth century—not only Marx but all the "conservatives" as well, such as J. P. Morgan, Bismarck, and Disraeli—practically all these gains have accrued to the blue-collar worker, half of the gains in the form of sharply reduced working hours (with the cuts ranging from 40 percent in Japan to 50 percent in Germany), half of them in the form of a twenty-fivefold increase in the real wages of blue-collar workers making or moving things.

There were thus very good reasons why the *rise* of blue-collar workers was peaceful rather than violent, let alone "revolutionary." But what explains that the *fall* of the blue-collar worker has been equally peaceful and almost entirely free of social protest, of upheaval, of serious dislocation, at least in the United States?

The Rise of the Knowledge Worker

The rise of the "class" succeeding the industrial blue-collar worker is not an opportunity to him. It is a challenge. The newly emerging dominant group are "knowledge workers." Knowledge workers amount to a third or more of the workforce in the United States, that is, to as large a proportion as industrial blue-collar workers ever were, except in wartime. The majority of knowledge workers are paid at least as well as blue-collar workers ever were, or better. And the new jobs offer much greater opportunities to the individual.

But—and it is a big but—the new jobs require, in the great majority, qualifications the blue-collar worker does not possess and is poorly equipped to acquire. The new jobs require a good deal of formal education and the ability to acquire and to apply theoretical and analytical knowledge. They require a different approach to work and a different mind-set. Above all, they require a habit of continual learning.

Displaced industrial workers thus cannot simply move into knowledge work or services work the way displaced farmers and displaced domestic workers moved into industrial work.

Even in communities that were totally dependent on one or two mass-production plants that have gone out of business or have cut employment by two-thirds—steel cities in western Pennsylvania or eastern Ohio, for instance, or car cities like Flint, Michigan—unemployment rates for adult, nonblack men and women fell within a few short years to levels barely higher than the U.S. average. And that means to levels barely higher than the U.S. "full-employment" rate. And there has been no radicalization of America's blue-collar workers.

The only explanation is that for the nonblack, blue-collar community the development came as no surprise, however unwelcome, painful, and threatening to individual worker and individual family. Psychologically—in terms of values perhaps, rather than in terms of emotions—America's industrial blue-collar workers must have been prepared to accept as right and proper the shift to jobs that require formal education and that pay for knowledge rather than for manual work, whether skilled or unskilled.

One possible factor may have been the GI Bill of Rights after World War II, which by offering a college education to every returning American veteran established advanced education as the "norm" and everything less as "substandard." Another factor may have been the draft the United States introduced in World War II and maintained for thirty-five years afterward, as a result of which the great majority of American male adults born between 1920 and

1950—and that means the majority of American adults alive today—served in the military for several years where they were *forced* to acquire a high-school education if they did not already have one. But whatever the explanation, in the United States the shift *to* knowledge work *from* blue-collar manual work making and moving things has largely been accepted (except in the black community) as appropriate or, at least, as inevitable.

In the United States the shift, by 1990 or so, had largely been accomplished. But so far only in the United States. In the other developed free-market countries, in western and northern Europe, and in Japan, it was just beginning in the 1990s. It is, however, certain to proceed rapidly in these countries from now on, and perhaps to proceed there faster than it originally did in the United States. Will it then also proceed, as it did by and large in the United States, with a minimum of social upheaval, of social dislocation, of social unrest? Or will the American development turn out to be another example of "American exceptionalism" (as has so much of American social history and especially of American labor history)? In Japan, the superiority of formal education and of the formally educated person is generally accepted so that the fall of the industrial worker—still a fairly recent class in Japan and outnumbering farmers and domestic servants only since well after World War II—may well be accepted as appropriate as it has been in the United States, and perhaps even more so. But what about industrialized Europe— the United Kingdom, Germany, France, Belgium, northern Italy, and so on—where there has been a "working-class culture" and a "self-respecting working class" for well over a century, and where, despite all evidence to the contrary, the belief is still deeply ingrained that industrial, blue-collar work, rather than knowledge, is the creator of all wealth? Will Europe react the way the American black has reacted? This surely is a key question, the answer to which will largely determine the social as well as the economic future of the developed free-market countries of Europe. And the answer will be given within the next decade or so.

The Emerging Knowledge Society

Knowledge workers will not be the majority in the knowledge society. But in many countries, if not most developed countries, they will be the largest single group in the population and the workforce. And even if outnumbered by other groups, knowledge workers will be the group that gives the emerging knowledge society its character, its leadership, its social profile. They may not be the *ruling* class of the knowledge society, but they already are its *leading* class. And in their characteristics, their social position, their values, and their expectations, they differ fundamentally from any group in history that has ever occupied the leading, let alone the dominant, position.

In the first place, the knowledge worker gains access to work, job, and social position through *formal education*.

The first implication of this is that education will become the center of the knowledge society, and schooling its key institution. What knowledge is required for everybody? What mix of knowledges is required for everybody? What is "quality" in learning and teaching? All these will, of necessity, become central concerns of the knowledge society, and central political issues. In fact, it may not be too fanciful to anticipate that the acquisition and distribution of formal knowledge will come to occupy the place in the politics of the knowledge society that acquisition and distribution of property and income have occupied in the two or three centuries that we have come to call the Age of Capitalism.

We can also predict with high probability that we will redefine what it means to be an "educated person."

The knowledge society will inevitably become *far more competitive* than any society we have yet known—for the simple reason that with knowledge being universally accessible, there are no excuses for nonperformance. There will be no "poor" countries. There will only be ignorant countries. And the same will be true for individual companies, individual industries, and individual organizations of any kind. It will be true for the individual, too. In fact, developed societies have already become infinitely more competitive for the individual than were the societies of the early twentieth century—let alone earlier societies, those of the nineteenth or eighteenth centuries. Then, most people had no opportunity to rise out of the "class" into which they were born, with most individuals following their fathers in their work and in their station in life.

But knowledge workers, whether their knowledge is primitive or advanced, whether they possess a little of it or a great deal, will, by definition, be specialized. Knowledge in application is effective only when it is specialized. Indeed, it is more effective, the more highly specialized it is.

Equally important is the second implication of the fact that knowledge workers are, of necessity, specialists: the need for them to work as members of an organization. It is only the organization that can provide the basic continuity that knowledge workers need to be effective. It is only the organization that can convert the specialized knowledge of the knowledge worker into performance.

By itself, specialized knowledge yields no performance. The surgeon is not effective unless there is a diagnosis, which, by and large, is not the surgeon's task and not even within the surgeon's competence. Market researchers, by themselves, produce only data. To convert the data into information, let alone to make them effective in knowledge action, requires marketing people, production people, service people. As a loner in his or her own research and writing, the historian can be very effective. But to produce the education of students, a great many other specialists have to contribute—people whose speciality may be literature, or mathematics, or other areas of history. And this requires that the specialist have access to an organization.

This access may be as a consultant. It may be as a provider of specialized services. But for a large number of knowledge workers, it will be as employees of an organization—full-time or part-time—whether a government agency, a hospital, a university, a business, a labor union, or any of hundreds of others. In the knowledge society, it is not the individual who performs. The individual is a cost center rather than a performance center. It is the organization that performs.

The Employee Society

The knowledge society is an *employee society.* Traditional society, that is, society before the rise of the manufacturing enterprise and the blue-collar manufacturing worker, was not a society of independents. Thomas Jefferson's society of independent small farmers, each being the owner of his own family farm and farming it without any help except for that of his wife and his children, was never much more than fantasy. Most people in history were dependents. But they did not work for an organization. They were working for an owner, as slaves, as serfs, as hired hands on the farm; as journeymen and apprentices in the craftsman's shop; as shop assistants and salespeople for a merchant; as domestic servants, free or unfree; and so on. They worked for a "master." When blue-collar work in manufacturing first arose, they still worked for a "master."

In Charles Dickens's great 1854 novel *Hard Times,* the workers work for an "owner." They do not work for the "factory." Only late in the nineteenth century did the factory rather than the owner become the employer. And only in the twentieth century did the corporation, rather than the factory, then become the employer. Only in this century has the "master" been replaced by a "boss," who, himself, ninety-nine times out of a hundred, is an employee and has a boss himself.

Knowledge workers will be both "employees" who have a "boss" and "bosses" who have "employees."

Organizations were not known to yesterday's social science, and are, by and large, not yet known to today's social science.

The first "organization" in the modern sense, the first that was seen as being prototypical rather than exceptional, was surely the modern business enterprise as it emerged after 1870—which is why, to this day, most people think of "management," as being "business management."

With the emergence of the knowledge society, we have become a society of organizations. Most of us work in and for an organization, are dependent for our effectiveness and equally for our living on access to an organization, whether as an organization's employee or as a provider of services to an organization—as a lawyer, for instance, or a freight forwarder. And more and more of these supporting services to organizations are, themselves, organized as organizations. The first law firm was organized in the United States a little over a century ago— until then lawyers had practiced as individuals. In Europe there were no law firms to speak of until after World War II. Today, the practice of law is increasingly done in larger and larger partnerships. But that is also true, especially in the United States, of the practice of medicine. The knowledge society is a society of organizations in which practically every social task is being performed in and through an organization.

Most knowledge workers will spend most if not all of their working life as "employees." But the meaning of the term is different from what it has been traditionally—and not only in English but in German, Spanish, and Japanese as well.

Individually, knowledge workers are dependent on the job. They receive a wage or salary. They are being hired and can be fired. Legally, each is an "employee." But collectively, they are the only "capitalists"; increasingly, through their pension funds and through their other savings (e.g., in the United States through mutual funds), the employees own the means of production. In traditional economics (and by no means only in Marxist economics), there is a sharp distinction between the "wage fund"—all of which went into consumption—and the "capital fund." And most social theory of industrial society is based, one way or another, on the relationship between the two, whether in conflict or in necessary and beneficial cooperation and balance. In the knowledge society, the two merge. The pension fund is "deferred wage" and, as such, a wage fund. But it is also increasingly the main source of capital, if not the only source of capital, for the knowledge society.

Equally important, and perhaps more important, is that in the knowledge society the employees, that is, knowledge workers, again own the tools of production. Marx's great insight was the realization that the factory worker does not and cannot own the tools of production and, therefore, has to be "alienated." There was no way, Marx pointed out, for workers to own the steam engine and to be able to take the steam engine with them when moving from one job to another. The capitalist had to own the steam engine and had to control it. Increasingly, the true investment in the knowledge society is not in machines and tools. It is in the knowledge worker. Without it, the machines, no matter how advanced and sophisticated, are unproductive.

The industrial worker needed the capitalist infinitely more than the capitalist needed the industrial worker—the basis for Marx's assertion that there would always be a surplus of industrial workers, and an "industrial reserve army" that would make sure that wages could not possibly rise above the subsistence level (probably Marx's most egregious error). In the knowledge society the most probable assumption—and certainly the assumption on which all organizations have to conduct their affairs—is that they need the knowledge worker far more than the knowledge worker needs them. It is up to the organization to market its knowledge jobs so as to obtain knowledge workers in adequate quantity and superior quality. The relationship increasingly is one of interdependence with the knowledge worker having to learn what the organization needs, but with the organization also having to learn what the knowledge worker needs, requires, and expects.

One additional conclusion: because the knowledge society perforce has to be a society of organizations, its central and distinctive organ is *management*.

When we first began to talk of management, the term meant "business management"—since large-scale business was the first of the new organizations to become visible. But we have learned this last half-century that management is the distinctive organ of all organizations. All of them require management—whether they use the term or not. All managers do the same things whatever the business of their organization. All of them have to bring people—each of them possessing a different knowledge—together for joint performance. All of them have to make human strengths productive in performance and human weaknesses irrelevant. All of them have to think through what are "results" in the organization—and have then to define objectives. All of them are responsible to think through what I call the "theory of the business," that is, the assumptions on which the organization bases its performance and actions, and equally, the assumptions that organizations make to decide what things not to do. All of them require an organ that thinks through strategies, that is, the means through which the goals of the organization become performance. All of them have to define the values of the organization, its system of rewards and punishments, and with it its spirit and its culture. In all of them, managers need both the knowledge of management as work and discipline and the knowledge and understanding of the organization itself, its purposes, its values, its environment and markets, its core competencies.

Management as a *practice* is very old. The most successful executive in all history was surely that Egyptian who, forty-seven hundred years or more ago, first conceived the pyramid—without any precedent—and designed and built it, and did so in record time. With a durability unlike that of any other human work, that first pyramid still stands. But as a *discipline,* management is barely fifty years old. It was first dimly perceived around the time of World War I. It did not emerge until World War II and then primarily in the United States. Since then, it has been the fastest-growing new business function, and its study the fastest-growing new academic discipline. No function in history has emerged as fast as management and managers have in the last fifty to sixty years, and surely none has had such worldwide sweep in such a short period.

Management, in most business schools, is still taught as a bundle of techniques, such as the technique of budgeting. To be sure, management, like any other work, has its own tools and its own techniques. But just as the essence of medicine is not the urinalysis, important though it is, the essence of management is not techniques and procedures. The essence of management is to make knowledge productive. Management, in other words, is a social function. And in its practice, management is truly a "liberal art."

The Social Sector

The old communities—family, village, parish, and so on—have all but disappeared in the knowledge society. Their place has largely been taken by the new unit of social integration: the organization. Where community membership was seen as fate, organization membership is voluntary. Where community claimed the entire person, organization is a means to a person's end, a tool. For two hundred years a hot debate has been raging, especially in the West: are communities "organic" or are they simply extensions of the person? Nobody would claim that the new organization is "organic." It is clearly an artifact, a human creation, a social technology.

But who, then, does the social tasks? Two hundred years ago social tasks were being done in all societies by the local community—primarily, of course, by the family. Very few, if any, of those tasks are now being done by the old communities. Nor would they be capable of doing them. People no longer stay where they were born, either in terms of geography or in terms of social position and status. By definition, a knowledge society is a society of mobility. And all the social functions of the old communities, whether performed well or poorly (and most were performed very poorly, indeed), presupposed that the individual and the family would stay put. Family is where they have to take you in, said a nineteenth-century adage; and community, to repeat, was fate. To leave the community meant becoming an outcast, perhaps even an outlaw. But the essence of a knowledge society is mobility in terms of where one lives, mobility in terms of what one does, mobility in terms of one's affiliation.

This very mobility means that in the knowledge society, social challenges and social tasks multiply. People no longer have "roots." People no longer have a "neighborhood" that controls where they live, what they do, and indeed, what their "problems" are allowed to be. The knowledge society, by definition, is a competitive society; with knowledge accessible to everyone, everyone is expected to place himself or herself, to improve himself or herself, and to have aspirations. It is a society in which many more people than ever before can be successful. But it is therefore, by definition, also a society in which many more people than ever before can fail, or at least can come in second. And if only because the application of knowledge to work has made developed societies so much richer than any earlier society could even dream of becoming, the failures, whether poverty or alcoholism, battered women or juvenile delinquents, are seen as failures of society. In traditional society they were taken for granted. In the knowledge society they are an affront, not just to the sense of justice but equally to the competence of society and its self-respect.

Who, then, in the knowledge society takes care of the social tasks? We can no longer ignore them. But traditional community is incapable of tackling them.

Two answers have emerged in this century—a majority answer and a dissenting opinion. Both have been proven to be the wrong answers.

The majority answer goes back more than a hundred years, to the 1880s, when Bismarck's Germany took the first faltering steps toward the welfare state. The answer: the problems of the social sector can, should, and must be solved by government. It is still probably the answer that most people accept, especially in the developed countries of the West—even though most people probably no longer fully believe it. But it has been totally disproven. Modern government, especially since World War II, has become a huge welfare bureaucracy everywhere. And the bulk of the budget in every developed country today is devoted to "entitlements," that is, to payment for all kinds of social services. And yet, in every developed country, society is becoming sicker rather than healthier, and social problems are multiplying. Government has a big role to play in social tasks—the role of policy-maker, of standard setter, and, to a substantial extent, the role of paymaster. But as the agency to *run* social services, it has proven itself almost totally incompetent—and we now know why.

The second dissenting opinion was first formulated by me in my 1942 book *The Future of Industrial Man*. I argued then that the new organization—and fifty years ago that meant the large business enterprise—would have to be the community in which the individual would find status and function, with the plant community becoming the place in and through which the social tasks would be organized. In Japan (though quite independently and without any debt to me) the large employer—government agency or business—has indeed increasingly attempted to become a "community" for its employees. "Lifetime employment" is only one affirmation of this. Company housing, company health plans, company vacations, and so on, all emphasize for the Japanese employee that the employer, and especially the big corporation, is the community and the successor to yesterday's village and to yesterday's family. But this, too, has not worked.

There is a need indeed, especially in the West, to bring the employee increasingly into the government of the plan community. What is now called "empowerment" is very similar to the things I talked about more than fifty years ago. But it does not create a community. And it does not create the structure through which the social tasks of the knowledge society can be tackled. In fact, practically all those tasks, whether providing education or health care; addressing the anomalies and diseases of a developed and, especially, of a rich society, such as alcohol and drug abuse; or tackling the problems of incompetence and irresponsibility such as those of the "underclass" in the American city—all lie outside the employing institution.

The employing institution is, and will remain, an "organization." The relationship between it and the individual is not that of "membership" in a "community," that is, an unbreakable, two-way bond.

To survive, it needs employment flexibility. But increasingly also, knowledge workers, and especially people of advanced knowledge, see the organization as the tool for the accomplishment of their own purposes and, therefore, resent—increasingly even in Japan—any attempt to subject them to the organization as a community, that is, to the control of the organization; to the demand of the organization that they commit themselves to lifetime membership; and to the demand that they subordinate their own aspirations to the goals and values of the organization. This is inevitable because the possessor of knowledge, as said earlier, owns his or her "tools of production" and has the freedom to move to wherever opportunities for effectiveness, for accomplishment, and for advancement seem greatest.

The right answer to the question, Who takes care of the social challenges of the knowledge society? is thus neither the government nor the employing organization. It is a separate and new *social sector.*

Increasingly, these organizations of the social sector serve a second and equally important purpose. *They create citizenship.* Modern society and modern polity have become so big and complex that citizenship, that is, responsible participation, is no longer possible. All we can do as citizens is to vote once every few years and to pay taxes all the time.

As a volunteer in the social sector institution, the individual can again make a difference.

Nothing has been disproved faster than the concept of the "organization man," which was almost universally accepted forty years ago. In fact, the more satisfying one's knowledge work is, the more one needs a separate sphere of community activity.

The New Pluralism

The emergence of the society of organizations challenges the function of government. All social tasks in the society of organizations are increasingly being done by individual organizations, each created for one, and only one, social task, whether education, health care, or street cleaning. Society, therefore, is rapidly becoming pluralist. Yet our social and political theories still assume a society in which there are no power centers except government. To destroy or at least to render impotent all other power centers was, in fact, the thrust of Western history and Western politics for five hundred years, from the fourteenth century on. It culminated in the eighteenth and nineteenth centuries when (except in the United States) such original institutions as still survived—for example, the universities or the established churches—all became organs of the state, with their functionaries becoming civil servants. But then, immediately beginning in the mid-nineteenth century, new centers arose—the first one, the modern business enterprise, emerged around 1870. And since then one new organization after another has come into being.

In the pluralism of yesterday, the feudalism of Europe's Middle Ages, or of Edo Japan in the seventeenth and eighteenth centuries, all pluralist organizations, whether a feudal baron in the England of the War of the Roses or the *daimyo*—the local lord—in Edo Japan, tried to be in control of whatever went on in their community. At least they tried to prevent anybody else from having control of any community concern or community institution within their domain.

But in the society of organizations, each of the new institutions is concerned only with its own purpose and mission. It does not claim power over anything else. But it also does not assume responsibility for anything else. *Who then is concerned with the common good?*

This has always been a central problem of pluralism. No earlier pluralism solved it. The problem is coming back now, but in a different guise. So far it has been seen as imposing limits on these institutions, that is, forbidding them to do things in the pursuit of their own mission, function, and interest that encroach upon the public domain or violate public policy. The laws against discrimination—by race, sex, age, education, health, and so on—that have proliferated in the United States in the last forty years all forbid socially undesirable behavior. But we are increasingly raising the question of the "social responsibility" of these institutions: What do these institutions have *to do*—in addition to discharging their own functions—to *advance* the public good? This, however—though nobody seems to realize it—is a demand to return to the old pluralism, the pluralism of feudalism. It is a demand for "private hands to assume public power."

That this could seriously threaten the functioning of the new organizations the example of the school in the United States makes abundantly clear.

The new pluralism has the old problem of pluralism—who takes care of the common good when the dominant institutions of society are single-purpose institutions? But it also has a new problem: how to maintain the performance capacity of the new institutions and yet maintain the cohesion of society? This makes doubly important the emergence of a strong and functioning social sector. It is an additional reason why the social sector will increasingly be crucial to the performance, if not to the cohesion, of the knowledge society.

As soon as knowledge became the key economic resource, the integration of the interests—and with it the integration of the pluralism of a modern polity—began to fall apart. Increasingly, noneconomic interests are becoming the new pluralism, the "special interests," the "single-cause" organizations, and so on. Increasingly, politics is not about "who gets what, when, how" but about values, each of them considered to be an absolute. Politics is about "the right to live" of the embryo in the womb as against the right of a woman to control her own body and to abort an embryo. It is about the environment. It is about gaining equality for groups alleged to be oppressed and discriminated against. None of these issues is economic. All are fundamentally moral.

Economic interests can be compromised, which is the great strength of basing politics on economic interests. "Half a loaf is still bread" is a meaningful saying. But "half a baby," in the biblical story of the judgment of Solomon, is not half a child. Half a baby is a corpse and a chunk of meat. There is no compromise possible. To an environmentalist, "half an endangered species" is an extinct species.

This greatly aggravates the crisis of modern government. Newspapers and commentators still tend to report in economic terms what goes on in Washington, in London, in Bonn, or in Tokyo. But more and more of the lobbyists who determine governmental laws and governmental actions are no longer lobbyists for economic interests. They lobby for and against measures they—and their paymasters—see as moral, spiritual, cultural. And each of these new moral concerns, each represented by a new organization, claims to stand for an absolute. Dividing their loaf is not compromising. It is treason.

There is thus in the society of organizations no single integrating force that pulls individual organizations in society and community into coalition. The traditional parties—perhaps the most successful political creations of the nineteenth century—no longer can integrate divergent groups and divergent points of view into a common pursuit of power. Rather, they become battlefields for these groups, each of them fighting for absolute victory and not content with anything but total surrender of the enemy.

This raises the question of how government can be made to function again. In countries with a tradition of a strong independent bureaucracy, notably Japan, Germany, and France, the civil service still tries to hold government together. But even in these countries the cohesion of government is increasingly being weakened by the special interests and, above all, by the noneconomic, the moral, special interests.

Since Machiavelli, almost five hundred years ago, political science has primarily concerned itself with power. Machiavelli—and political scientists and politicians since him—took it for granted that government can function once it has power. Now, increasingly, the questions to be tackled are: What are the functions that government and only government can discharge and that government *must* discharge? and How can government be organized so that it can discharge those functions in a society of organizations?

The twenty-first century will surely be one of continuing social, economic, and political turmoil and challenge, at least in its early decades. The Age of Social Transformations is not over yet. And the challenges looming ahead may be more serious and more daunting still than those posed by the social transformations of the twentieth century that have already happened.

Yet we will not even have a chance to resolve these new and looming problems of tomorrow unless we *first* address the challenges posed by the developments that are already accomplished facts. If the twentieth century was one of social transformations, the twenty-first century needs to be one of social and political innovations.

24.

THE COMING OF
ENTREPRENEURIAL SOCIETY

"Every generation needs a new revolution," was Thomas Jefferson's conclusion toward the end of his long life. His contemporary, Goethe, the great German poet, though an archconservative, voiced the same sentiment when he sang in his old age: *"Vernunft wird Unsinn/Wohltat, Plage."* (Reason becomes nonsense/Boons afflictions.)

Both Jefferson and Goethe were expressing their generation's disenchantment with the legacy of Enlightenment and French Revolution. But they might just as well have reflected on our present-day legacy, 150 years later, of that great shining promise, the welfare state, begun in Imperial Germany for the truly indigent and disabled, which has now become "everybody's entitlement" and an increasing burden on those who produce. Institutions, systems, policies, eventually outlive themselves, as do products, processes, and services. They do it when they accomplish their objectives and they do it when they fail to accomplish their objectives. The mechanisms may still tick. But the assumptions on which they were designed have become invalid—as, for example, have the demographic assumptions on which health-care plans and retirement schemes were designed in all developed countries over the last hundred years. Then, indeed, reason becomes nonsense and boons afflictions.

Yet "revolutions," as we have learned since Jefferson's days, are not the remedy. They cannot be predicted, directed, or controlled. They bring to power the wrong people. Worst of all, their results—predictably—are the exact opposite of their promises. Only a few years after Jefferson's death in 1826, that great anatomist of government and politics, Alexis de Tocqueville, pointed out that revolutions do not demolish the prisons of the old regime; they enlarge them. The most lasting legacy of the French Revolution, Tocqueville proved, was the tightening of the very fetters of pre-Revolutionary France: the subjection of the whole country to an uncontrolled and uncontrollable bureaucracy, and the centralization in Paris of all political, intellectual, artistic, and economic life. The main consequences of the Russian Revolution were new serfdom for the tillers of the land, an omnipotent secret police, and a rigid, corrupt, stifling bureaucracy—the very features of the czarist regime against which Russian liberals and revolutionaries had protested most loudly and with most justification. And the same must be said of Mao's macabre "Great Cultural Revolution."

Indeed, we now know that "revolution" is a delusion, the pervasive delusion of the nineteenth century, but today perhaps the most discredited of its myths. We now know that "revolution" is not achievement and the new dawn. It results from senile decay, from the bankruptcy of ideas and institutions, from failure of self-renewal.

And yet we also know that theories, values, and all the artifacts of human minds and human hands do age and rigidify, becoming obsolete, becoming "afflictions."

Innovation and entrepreneurship are thus needed in society as much as in the economy, in public-service institutions as much as in businesses. It is precisely because innovation and entrepreneurship are not "root and branch" but "one step at a time," a product here, a policy there, a public service yonder; because they are not planned but focused on this opportunity and that need; because they are tentative and will disappear if they do not produce the expected and needed results; because, in other words, they are pragmatic rather than dogmatic and modest rather than grandiose—that they promise to keep any society, economy, industry, public service, or business flexible and self-renewing. They achieve what Jefferson hoped to achieve through revolution in every generation, and they do so without bloodshed, civil war, or concentration camps, without economic catastrophe, but with purpose, with direction, and under control.

What we need is an entrepreneurial society in which innovation and entrepreneurship are normal, steady, and continual. Just as management has become the specific organ of all contemporary institutions, and the integrating organ of our society of organizations, so innovation and entrepreneurship have to become an integral life-sustaining activity in our organizations, our economy, our society.

This requires of executives in all institutions that they make innovation and entrepreneurship a normal, ongoing, everyday activity, a practice in their own work and in that of their organization.

Planning Does Not Work

The first priority in talking about the public policies and governmental measures needed in the entrepreneurial society is to define what will not work—especially as the policies that will not work are so popular today.

"Planning" as the term is commonly understood is actually incompatible with an entrepreneurial society and economy. Innovation does indeed need to be purposeful and entrepreneurship has to be managed. But innovation, almost by definition, has to be decentralized, ad hoc, autonomous, specific, and microeconomic. It had better start small, tentative, flexible. Indeed, the opportunities for innovation are found, on the whole, only way down and close to events. They are not to be found in the massive aggregates with which the planner deals of necessity, but in the deviations therefrom—in the unexpected, in the incongruity, in the difference between "the glass half full" and "the glass half empty," in the weak link in a process. By the time the deviation becomes "statistically significant" and thereby visible to the planner, it is too late. Innovative opportunities do not come with the tempest but with the rustling of the breeze.

Systematic Abandonment

One of the fundamental changes in worldview and perception of the last twenty years—a truly monumental turn—is the realization that governmental policies and agencies are of human rather than of divine origin, and that therefore the one thing certain about them is that they will become obsolete fairly fast. Yet politics is still based on the age-old assumption that whatever government does is grounded in the nature of human society and therefore "forever." As a result no political mechanism has so far arisen to slough off the old, the outworn, the no-longer-productive in government.

Or rather, what we have is not working yet. In the United States there has lately been a rash of "sunset laws," which prescribe that a governmental agency or a public law lapse after a certain period of time unless specifically reenacted. These laws have not worked, however—in part because there are no objective criteria to determine when an agency or a law becomes dysfunctional, in part because there is so far no organized process of abandonment, but perhaps mostly because we have not yet learned to develop new or alternative methods for achieving what an ineffectual law or agency was originally supposed to achieve. To develop both the principles and the process for making "sunset laws" meaningful and effective is one of the important social innovations ahead of us—and one that needs to be made soon. Our societies are ready for it.

A Challenge for the Individuals

In an entrepreneurial society individuals face a tremendous challenge, a challenge they need to exploit as an opportunity: the need for continual learning and relearning.

In traditional society it could be assumed—and was assumed—that learning came to an end with adolescence or, at the latest, with adulthood. What one had not learned by age twenty-one or so, one would never learn. But also what one had learned by age twenty-one or so one would apply, unchanged, the rest of one's life. On these assumptions was based traditional apprenticeship, traditional crafts, traditional professions, but also the traditional systems of education and the schools. Crafts, professions, systems of education, and schools are still, by and large, based on these assumptions. There were, of course, always exceptions, some groups that practiced continual learning and relearning: the great artists and the great scholars, Zen monks, mystics, the Jesuits. But these exceptions were so few that they could safely be ignored.

In an entrepreneurial society, however, these "exceptions" become the exemplars. The correct assumption in an entrepreneurial society is that individuals will have to learn new things well after they have become adults—and maybe more than once. The correct assumption is that what individuals have learned by age twenty-one will begin to become obsolete five to ten years later and will have to be replaced—or at least refurbished—by new learning, new skills, new knowledge.

One implication of this is that individuals will increasingly have to take responsibility for their own continual learning and relearning, for their own self-development and for their own careers. They can no longer assume that what they have learned as children and youngsters will be the "foundation" for the rest of their lives. It will be the "launching pad"—the place to take off from rather than the place to build on and to rest on. They can no longer assume that they "enter upon a career" that then proceeds along a predetermined, well-mapped, and well-lighted "career path" to a known destination—what the American military calls "progressing in grade." The assumption from now on has to be that individuals on their own will have to find, determine, and develop a number of "careers" during their working lives.

And the more highly schooled the individuals, the more entrepreneurial their careers and the more demanding their learning challenges. The carpenter can still assume, perhaps, that the skills he acquired as apprentice and journeyman will serve him forty years later. Physicians, engineers, metallurgists, chemists, accountants, lawyers, teachers, managers, had better assume that the skills, knowledges, and tools they will have to master and apply fifteen years hence are going to be different and new. Indeed, they better assume that fifteen years hence they will be doing new and quite different things, will have new and different goals and, indeed, in many cases, different "careers." And only they themselves can take responsibility for the necessary learning and relearning, and for directing themselves. Tradition, convention, and "corporate policy" will be a hindrance rather than a help.

This also means that an entrepreneurial society challenges habits and assumptions of schooling and learning. The educational systems the world over are in the main extensions of what Europe developed in the seventeenth century. There have been substantial additions and modifications. But the basic architectural plan on which our schools and universities are built goes back three hundred years and more. Now new, in some cases radically new, thinking and new, in some cases radically new, approaches are required, and on all levels.

Using computers in preschool may turn out to be a passing fad. But four-year-olds exposed to television expect, demand, and respond to very different pedagogy than four-year-olds did fifty years ago.

Young people headed for a "profession"—that is, four-fifths of today's college students—do need a "liberal education." But that clearly means something quite different from the nineteenth-century version of the seventeenth-century curriculum that passed for a "liberal education" in the English-speaking world or for *"Allgemeine Bildung"* in Germany. If this challenge is not faced up to, we risk losing the fundamental concept of a "liberal education" altogether and will descend into the purely vocational, purely specialized, which would endanger the educational foundation of the community and, in the end, community itself. But also educators will have to accept that schooling is not for the young only and that the greatest challenge—but also the greatest opportunity—for the school is the continuing relearning of already highly schooled adults.

So far we have no educational theory for these tasks.

So far we have no one who does what, in the seventeenth century, the great Czech educational reformer Johann Comenius did or what the Jesuit educators did when they developed what to this day is the "modern" school and the "modern" university.

But in the United States, at least, practice is far ahead of theory. To me the most positive development in the last twenty years, and the most encouraging one, is the ferment of educational experimentation in the United States—a happy by-product of the absence of a "Ministry of Education"—in respect to the continuing learning and relearning of adults, and especially of highly schooled professionals. Without a "master plan," without "educational philosophy," and, indeed, without much support from the educational establishment, the continuing education and professional development of already highly educated and highly achieving adults has become the true "growth industry" in the United States in the last twenty years.

The emergence of the entrepreneurial society may be a major turning point in history.

A hundred years ago the worldwide panic of 1873 terminated the century of laissez-faire that had begun with the publication of Adam Smith's *Wealth of Nations* in 1776. In the panic of 1873 the modern welfare state was born. A hundred years later it had run its course, almost everyone now knows. It may survive despite the demographic challenges of an aging population and a shrinking birthrate. But it will survive only if the entrepreneurial economy succeeds in greatly raising productivities. We may even still make a few minor additions to the welfare edifice, put on a room here or a new benefit there. But the welfare state is past rather than future—as even the old liberals now know.

Will its successor be the entrepreneurial society?

25.

CITIZENSHIP THROUGH
THE SOCIAL SECTOR

S ocial needs will grow in two areas. They will grow, first, in what
has traditionally been considered *charity:* helping the poor, the
disabled, the helpless, the victims. And they will grow, perhaps even
faster, in respect to services that aim at *changing the community* and
at *changing people.*

In a transition period, the number of people in need always
grows. There are the huge masses of refugees all over the globe, victims of war and social upheaval, of racial, ethnic, political, and religious persecution, of government incompetence and of government
cruelty. Even in the most settled and stable societies people will be
left behind in the shift to knowledge work. It takes a generation or
two before a society and its population catch up with radical
changes in the composition of the workforce and in the demands
for skills and knowledge. It takes some time—the best part of a generation, judging by historical experience—before the productivity
of service workers can be raised sufficiently to provide them with a
"middle-class" standard of living.

The needs will grow equally—perhaps even faster—in the second area of social services, services that do not dispense charity but

attempt to make a difference in the community and to change people. Such services were practically unknown in earlier times, whereas charity has been with us for millennia. But they have mushroomed in the last hundred years, especially in the United States.

These services will be needed even more urgently in the next decades. One reason is the rapid increase in the number of elderly people in all developed countries, many of whom live alone and want to live alone. A second reason is the growing sophistication of health care and medical care, calling for health-care research, health-care education, and for more and more medical and hospital facilities. Then there is the growing need for continuing education of adults, and the need created by the growing number of one-parent families. The community-service sector is likely to be one of the true "growth sectors" of developed economies, whereas we can hope that the need for charity will eventually subside again.

A "Third Sector"

None of the U.S. programs of the last forty years in which we tried to tackle a social problem through government action has produced significant results. But independent nonprofit agencies *have* had impressive results. Public schools in inner cities—for example, New York, Detroit, and Chicago—have been going downhill at an alarming rate. Church-run schools (especially schools of the Roman Catholic dioceses) have had startling successes—in the same communities, and with children from similarly broken families and of similar racial and ethnic groups. The only successes in fighting alcoholism and drug abuse (very substantial ones) have been achieved by such independent organizations as Alcoholics Anonymous, the Salvation Army, and the Samaritans. The only successes in getting "welfare mothers"—single mothers, often black or Hispanic—off welfare and back into paid work and a stable family life have been achieved by autonomous, nonprofit organizations such as the Judson Center in Royal Oak, Michigan. Improvements in major health-care areas such as the prevention and treatment of cardiac disease and of mental illness have largely been the work of independent nonprofit organizations. The American Heart Association and the American Mental Health Association, for instance, have sponsored the necessary research and taken the lead in educating both the medical community and the public in prevention and treatment.

To foster autonomous community organizations in the social sector is therefore an important step in turning government around and making it perform again.

But the greatest contribution that the autonomous community organization makes is as a new *center of meaningful citizenship*. The megastate has all but destroyed citizenship. To restore it, the postcapitalist polity needs a "third sector," in addition to the two generally recognized ones, the "private sector" of business and the "public sector" of government. It needs an autonomous *social sector*.

In the megastate, political citizenship can no longer function. Even if the country is small, the affairs of government are so far away that individuals cannot make a difference.

Individuals can vote—and we have learned the hard way these last decades how important the right to vote is. Individuals can pay taxes—and again we have learned the hard way these last decades that this is a meaningful obligation.

The individuals cannot take responsibility, cannot take action to make a difference. Without citizenship, however, the polity is empty. There can be nationalism, but without citizenship, it is likely to degenerate from patriotism into chauvinism. Without citizenship, there cannot be that responsible commitment that creates the citizen and that in the last analysis holds together the body politic. Nor can there be the sense of satisfaction and pride that comes from making a difference. Without citizenship, the political unit, whether called "state" or "empire," can only be a power. Power is then the only thing that holds it together. In order to be able to act in a rapidly changing and dangerous world, the postcapitalist polity must re-create citizenship.

The Need for Community

Equally, there is a need to restore community. Traditional communities no longer have much integrating power; they cannot survive the mobility that knowledge confers on the individual. Traditional communities, we have now learned, were held together far less by what their members had in common than by necessity, if not by coercion and fear.

The traditional family was a necessity. In nineteenth-century fiction most families were what we would now call "broken families." But they had to stay together, no matter how great their hatred, their loathing, their fear of each other. Family is where they have to take you in, was a nineteenth-century saying. Family before this century provided practically all the social services available.

To cling to family was a necessity; to be repudiated by it, a catastrophe. A stock figure of American plays and movies as late as the 1920s was the cruel father who threw out his daughter when she came home with an illegitimate child. She then had only two choices: to commit suicide or to become a prostitute.

Today, family is actually becoming more important to most people. But it is becoming so as a voluntary bond, as a bond of affection, of attachment, of mutual respect, rather than one of necessity. Today's young people, once they have grown out of adolescent rebellion, feel a much greater need than my generation did to be close to their parents and to their siblings.

Still, family no longer makes up the community. But people do need a community. They need it particularly in the sprawling huge cities and suburbs in which more and more of us live. One can no longer count—as one could in the rural village—on neighbors who share the same interests, the same occupations, the same ignorance, and who live together in the same world. Even if the bond is close, one cannot count on family. Geographic and occupational mobility mean that people no longer stay in the place, class, or culture where they were born, where their parents live, where their siblings and their cousins live. The community that is needed in postcapitalist society—and needed especially by the knowledge worker—has to be based on *commitment and compassion* rather than being imposed by proximity and isolation.

Forty years ago, I thought that this community would come into being at the place of work. In *The Future of Industrial Man* (1942), *The New Society* (1949), and *The Practice of Management* (1954), I talked of the plant community as the place that would grant the individual status and function, as well as the responsibility of self-government. But even in Japan, the plant community is not going to work much longer. It is becoming increasingly clear that the Japanese plant community is based far less on a sense of belonging than on fear. A worker in a large Japanese company with its seniority wage system who loses his job past age thirty has become virtually unemployable for the rest of his life.

In the West, the plant community never took root. I still strongly maintain that the employee has to be given the maximum responsibility and self-control—the idea that underlay my advocacy of the plant community. The knowledge-based organization has to become a responsibility-based organization.

But individuals, and especially knowledge workers, need an additional sphere of social life, of personal relationships, and of contribution outside and beyond the job, outside and beyond the organization, indeed, outside and beyond their own specialized knowledge area.

The Volunteer as Citizen

The one area in which this need can be satisfied is the social sector. There, individuals can contribute. They can have responsibility. They can make a difference. They can be "volunteers."

This is already happening in the United States.

The denominational diversity of American churches; the strong emphasis on local autonomy of states, counties, cities; and the community tradition of isolated frontier settlements all slowed down the politicization and centralization of social activities in the United States. As a result, America now has almost one million nonprofit organizations active in the social sector. They represent as much as one-tenth of the gross national product—one-quarter of that sum raised by donations from the public, another quarter paid by government for specific work (e.g., to administer health-care reimbursement programs), the rest fees for services rendered (e.g., tuition paid by students attending private universities or money made by the art stores to be found now in every American museum).

The nonprofits have become America's biggest employer. Every other American adult (90 million people all told) works at least three hours a week as "unpaid staff," that is, as a volunteer for a nonprofit organization, for churches and hospitals; for health-care agencies, for community services like Red Cross, Boy Scouts, and Girl Scouts; for rehabilitation services like Salvation Army and Alcoholics Anonymous; for shelters for battered wives; and for tutoring services in inner-city schools. By the year 2000 or 2010, the number of such unpaid staff people should have risen to 120 million, and their average hours of work to five per week.

These volunteers are no longer "helpers"; they have become "partners." Nonprofit organizations in the United States increasingly have a full-time paid executive, but the rest of the management team are volunteers. Increasingly, they run the organization.

The greatest change has taken place in the American Catholic Church. In one major diocese, lay women now actually run all the parishes as "parish administrators." The priests say mass and dispense the sacraments. Everything else, including all the social and community work of the parishes, is done by "unpaid staff," led by the parish administrator.

The main reason for this upsurge of volunteer participation in the United States is not an increase in need. The main reason is the search on the part of the volunteers for community, for commitment, for contribution. The great bulk of the new volunteers are not retired people; they are husbands and wives in the professional, two-earner family, people in their thirties and forties, well educated, affluent, busy. They enjoy their jobs. But they feel the need to do something where "we can make a difference," to use the phrase one hears again and again—whether that means running a Bible class in the local church; teaching disadvantaged children the multiplication tables; or visiting old people back home from a long stay in the hospital and helping them with their rehabilitation exercises.

What the U.S. nonprofits do for their volunteers may well be just as important as what they do for the recipients of their services.

The Girl Scouts of America is one of the few American organizations that has become racially integrated. In their troops, girls regardless of color or national origin work together and play together. But the greatest contribution of the integration drive that the Girl Scouts began in the 1970s is that it recruited a large number of mothers—Black, Asian, Hispanic—into leadership positions as volunteers in integrated community work.

Citizenship in and through the social sector is not a panacea for the ills of postcapitalist society and postcapitalist polity, but it may be a prerequisite for tackling these ills. It restores the civic responsibility that is the mark of citizenship, and the civic pride that is the mark of community.

The need is greatest where community and community organizations—and citizenship altogether—have been so thoroughly damaged as to have been almost totally destroyed: in the ex-Communist countries. Government in these countries has not only been totally discredited; it has become totally impotent. It may take years before the successor governments to the Communists—in Czechoslovakia and Kazakhstan, in Russia, Poland, and Ukraine—can competently carry out the tasks that only government can do: managing money and taxes; running the military and the courts; conducting foreign relations. In the meantime, only autonomous, local nonprofits—organizations of the social sector based on volunteers and releasing the spiritual energies of people—can provide both the social services that the society needs and the leadership development that the polity needs.

Different societies and different countries will surely structure the social sector very differently. But every developed country needs an autonomous, self-governing social sector of community organizations—to provide the requisite community services, but above all to restore the bonds of community and a sense of active citizenship. Historically, community was fate. In the postcapitalist society and polity, community has to become commitment.

26.

FROM ANALYSIS TO PERCEPTION—THE NEW WORLDVIEW

Around 1680 a French physicist, Denis Papin, then working in Germany—as a Protestant he had been forced to leave his native country—invented the steam engine. Whether he actually built one we do not know; but he designed one, and he actually put together the first safety valve. A generation later, in 1712, Thomas Newcomen then put the first working steam engine into an English coal mine. This made it possible for coal to be mined—until then groundwater had always flooded English mines. With Newcomen's engine, the Age of Steam was on. Thereafter, for 250 years, the model of technology was mechanical. Fossil fuels rapidly became the main source of energy. And the ultimate source of motive power was what happens inside a star, that is, the sun. In 1945, atomic fission and, a few years later, fusion replicated what occurs in the sun. There is no going beyond this. In 1945, the era in which the mechanical universe was the model came to an end. Just a year later, in 1946, the first computer, the ENIAC, came on stream. And with it began an age in which information will be the organizing principle for work. Information, however, is the basic principle of biological rather than of mechanical processes.

Very few events have as much impact on civilization as a change in the basic principle for organizing work. Up until A.D. 800 or 900, China was far ahead of any Western country in technology, in science, and in culture and civilization generally. Then the Benedictine monks in northern Europe found new sources of energy. Until that point the main source of energy, if not the only one, had been a two-legged animal called man. It was the peasant's wife who pulled the plow. The horse collar for the first time made it possible to replace the farmer's wife with animal power. And the Benedictines also converted what in antiquity were toys, waterwheel and windmill, into the first machines. Within two hundred years technological leadership shifted from China to the Occident. Seven hundred years later Papin's steam engine created a new technology and with it a new worldview—the mechanical universe.

In 1946, with the advent of the computer, information became the organizing principle of production. With this, a new basic civilization came into being.

The Social Impacts of Information

A great deal these days (almost too much) is being said and written about the impact of the information technologies on the material civilization, on goods, services, and businesses. The social impacts are, however, as important; indeed, they may be more important. One of the impacts is widely noticed: any such change triggers an explosion of entrepreneurship. In fact, the entrepreneurial surge that began in the United States in the late 1970s, and which within ten years had spread to all non-Communist developed countries, is the fourth such surge since Denis Papin's time three hundred years ago. The first one ran from the middle of the seventeenth century through the early years of the eighteenth century; it was triggered by the "Commercial Revolution," the tremendous expansion of trade following the development of the first oceangoing freighter that could actually carry heavy payloads over large distances. The second entrepreneurial surge—beginning in the middle of the eighteenth century and running to the middle of the nineteenth—was what we commonly call the Industrial Revolution. Then, around 1870, the third entrepreneurial surge was triggered by the new industries— the first ones that did not just apply different motive power but actually turned out products that had never been made before or only in minute quantities: electricity, telephone, electronics, steel, chemicals and pharmaceuticals, automobiles and airplanes.

We are now in a fourth surge, triggered by information and biology. Like the earlier entrepreneurial surges, the present one is not confined to "high tech"; it embraces equally "middle tech," "low tech," and "no tech." Like the earlier ones, it is not confined to new or small enterprises, but is carried by existing and big ones as well—and often with the greatest impact and effectiveness. And, like the earlier surges, it is not confined to "inventions," that is, to technology. Social innovations are equally "entrepreneurial" and equally important. Some of the social innovations of the Industrial Revolution—the modern army, the civil service, the postal service, the commercial bank—have surely had as much impact as railroad or steamship. Similarly, the present age of entrepreneurship will be as important for its social innovations—and especially for innovations in politics, government, education, and economics—as for any new technology or material product.

Another important social impact of information is also visible and widely discussed: the impact on the national state and, particularly, on that twentieth-century hypertrophy of the national state, the totalitarian regime. Itself a creature of the modern media, newspapers, movies, and radio, it can exist only if it has total control of information. But with everyone being able to receive information directly from a satellite in the home—and on "dishes" already so small that no secret police can hope to find them—control of information by government is no longer possible. Indeed, information is now transnational; like money, information has no "fatherland."

Since information knows no national boundaries, it will also form new "transnational" communities of people who, maybe without ever seeing each other in the flesh, are in communion because they are in communication. The world economy, especially the "symbol economy" of money and credit, is already one of the non-national, transnational communities.

Other social impacts are just as important but rarely seen or discussed. One of them is the likely transformation of the twentieth-century city. Today's city was created by the great breakthrough of the nineteenth century: the ability to move people to work by means of train and streetcar, bicycle and automobile. It will be transformed by the great twentieth-century breakthrough: the ability to move work to people by moving ideas and information. In fact, the city—central Tokyo, central New York, central Los Angeles, central London, central Paris, central Bombay—has already outlived its usefulness. We no longer can move people into and out of it, as witness the two-hour trips in packed railroad cars to reach the Tokyo or New York office building, the chaos in London's Piccadilly Circus, or the two-hour traffic jams on the Los Angeles freeways every morning and evening. We are already beginning to move the information to where the people are—outside the cities—in such work as the handling of credit cards, of engineering designs, of insurance policies and insurance claims, or of medical records. Increasingly, people will work in their homes or, as many more are likely to do, in small "office satellites" outside the crowded central city. The facsimile machine, the telephone, the two-way video screen, the telex, the teleconference, are taking over from railroad, automobile, and from airplane as well. The real-estate boom in all the big cities in the 1970s and '80s, and the attendant skyscraper explosion, are not signs of health. They signal the beginning of the end of the central city. The decline may be slow; but we no longer need that great achievement, the central city, at least not in its present form.

The city might become an information center rather than a center for work—the place from which information (news, data, music) radiates. It might resemble the medieval cathedral where the peasants from the surrounding countryside congregated once or twice a year at the great feast days; in between, it stood empty except for its learned clerics and its cathedral school. And will tomorrow's university be a "knowledge center" that transmits information, rather than a place that students actually attend?

Where work is done determines in large measure also how it is done. It strongly affects what work is being done. That there will be great changes we can be certain—but how and when so far we cannot even guess.

Form and Function

The question of the right size for a given task or a given organization will become a central challenge. Greater performance in a mechanical system is obtained by scaling up. Greater power means greater output: bigger is better. But this does not hold for biological systems. There size follows function.

It would surely be counterproductive for the cockroach to be big, and equally counterproductive for the elephant to be small. As biologists are fond of saying, The rat knows everything it needs to be successful as a rat. Whether the rat is more intelligent than the human being is a stupid question; in what it takes to be successful as a rat, the rat is way ahead of any other animal, including the human being. In an information-based society, bigness becomes a "function" and a dependent, rather than an independent, variable. In fact, the characteristics of information imply that the smallest effective size will be best. "Bigger" will be "better" only if the task cannot be done otherwise.

For communication to be effective, there has to be both information and meaning. And meaning requires communion. If somebody whose language I do not speak calls me on the telephone, it doesn't help me at all that the connection is crystal clear. There is no "meaning" unless I understand the language—the message the meteorologist understands perfectly is gibberish to a chemist. Communion, however, does not work well if the group is very large. It requires constant reaffirmation. It requires the ability to interpret. It requires a community. "I know what this message means because I know how our people think in Tokyo, or in London, or in Beijing." *I know* is the catalyst that converts "information" into "communications."

For fifty years, from the early days of the Great Depression to the 1970s, the trend ran toward centralization and bigness. Prior to 1929, doctors did not put their paying patients into hospitals except for surgery. Very few babies before the 1920s were born in hospitals; the majority were born at home. The dynamics of higher education in the United States as late as the 1930s were in the small and medium-size liberal arts colleges. After World War II, they shifted increasingly to the big university and to the even bigger "research university." The same thing happened in government. And after World War II, bigness became an obsession in business. Every firm had to be a "billion-dollar corporation."

In the 1970s the tide turned. No longer is it the mark of good government to be bigger. In health care we now assert that whatever can be done outside the hospital better be done elsewhere. Before the 1970s, even mildly sick mental patients in the United States were considered to be best off in a mental hospital. Since then, mental patients who are no threat to others have been pushed out of the hospital (not always with good results). We have moved away from the worship of size that characterized the first three quarters of the century and especially the immediate post–World War II period. We are rapidly restructuring and "divesting" big business. We are, especially in the United States, pushing governmental tasks away from the center and toward local government in the country. We are "privatizing" and farming out governmental tasks, especially in the local community, to small outside contractors.

Increasingly, therefore, the question of the right size for a task will become a central one. Is this task best done by a bee, a hummingbird, a mouse, a deer, or an elephant? All of them are needed, but each for a different task and in a different ecology. The right size will increasingly be whatever handles most effectively the information needed for task and function. Where the traditional organization was held together by command and control, the "skeleton" of the information-based organization will be the optimal information system.

From Analysis to Perception

Technology is not nature, but humanity. It is not about tools; it is about how people work. It is equally about how they live and how they think. There is a saying of Alfred Russel Wallace, the co-discoverer—with Charles Darwin—of the theory of evolution: "Man is the only animal capable of directed and purposeful evolution; he makes tools." But precisely because technology is an extension of human beings, basic technological change always both expresses our worldview and, in turn, changes it.

The computer is in one way the ultimate expression of the analytical, the conceptual worldview of a mechanical universe that arose in Denis Papin's time, the late seventeenth century. It rests, in the last analysis, on the discovery of Papin's contemporary and friend, the philosopher-mathematician Gottfried Leibniz, that all numbers can be expressed "digitally," that is, by 1 and 0. It became possible because of the extension of this analysis beyond numbers to logic in Bertrand Russell and Alfred N. Whitehead's *Principia Mathematica* (published from 1910 through 1913), which showed that any concept can be expressed by 1 and 0 if made unambiguous and into "data."

But while it is the triumph of the analytical and conceptual model that goes back to Papin's own master, René Descartes, the computer also forces us to transcend that model. "Information" itself is indeed analytical and conceptual. But information is the organizing principle of every biological process. Life, modern biology teaches, is embodied in a "genetic code," that is, in programmed information. Indeed, the sole definition of that mysterious reality "life" that does not invoke the supernatural is that it is matter organized by information. And biological process is not analytical. In a mechanical phenomenon the whole is equal to the sum of its parts and therefore capable of being understood by analysis. Biological phenomena are, however, "wholes." They are different from the sum of their parts. Information is indeed conceptual. But meaning is not; it is perception.

In the worldview of the mathematicians and philosophers, which Denis Papin and his contemporaries formulated, perception was "intuition" and either spurious or mystical, elusive, mysterious. Science did not deny its existence (though a good many scientists did). It denied its validity. "Intuition," the analysts asserted, can neither be taught nor trained. Perception, the mechanical worldview asserts, is not "serious" but is relegated to the "finer things of life," the things we can do without. We teach "art appreciation" in our schools as indulgence in pleasure. We do not teach art as the rigorous, demanding discipline it is for the artist.

In the biological universe, however, perception is at the center. And it can—indeed it must—be trained and developed. We do not hear "C" "A" "T"; we hear "cat." "C" "A" "T" are "bits," to use the modern idiom; they are analysis. Indeed, the computer cannot do anything that requires meaning unless it goes beyond bits. That is what "expert systems" are about; they attempt to put into the logic of the computer, into an analytical process, the perception of experience that comes from understanding the whole of a task or subject matter.

In fact, we had begun to shift toward perception well before the computer. Almost a century ago, in the 1890s, configuration (Gestalt) psychology first realized that we hear "cat" and not "C" "A" "T." It first realized that we perceive. Since then almost all psychology—whether developmental, behavioral, or clinical—has shifted from analysis to perception. Even post-Freudian "psychoanalysis" is becoming "psychoperception" and attempts to understand the person rather than his or her mechanisms, the "drives." In governmental and business planning, we increasingly talk of "scenarios" in which a perception is the starting point. And, of course, any "ecology" is perception rather than analysis. In an ecology, the "whole" has to be seen and understood, and the "parts" exist only in contemplation of the whole.

When some fifty years ago the first American college—Bennington in Vermont—began to teach the doing of art—painting, sculpture, ceramics, playing an instrument—as integral parts of a liberal arts education, it was a brazen, heretical innovation that defied all respectable academic conventions. Today, every American college does this. Forty years ago the public universally rejected nonobjective modern painting. Now the museums and galleries showing the works of modern painters are crowded and their works fetch record prices. What is "modern" about modern painting is that it attempts to present what the painter sees rather than what the viewer sees. It is meaning rather than description.

Three hundred years ago, Descartes said, "I *think* therefore I am." We will now have to say also, "I *see* therefore I am." Since Descartes, the accent has been on the conceptual. Increasingly, we will balance the conceptual and the perceptual. Indeed, the new realities are *configurations* and as such call for perception as much as for analysis: the dynamic disequilibrium of the new pluralisms, for instance; the multitiered transnational economy and the transnational ecology; the new archetype of the "educated person" that is so badly needed. And *The New Realities* attempts as much to make us *see* as it attempts to make us *think*.

It took more than a hundred years after Descartes and his contemporary, Galileo, had laid the foundations for the science of the mechanical universe, until Immanuel Kant produced the metaphysics that codified the new worldview. His *Kritik der reinen Vernunft* (*Critique of Pure Reason,* 1781) then dominated Western philosophy for more than a century. It defined the meaningful questions even for Kant's adversaries, such as Friedrich Nietzsche. Indeed, Kant still defined "knowledge" even for Ludwig Wittgenstein in the first half of the twentieth century. But contemporary philosophers no longer focus on Kant's concerns. They deal with configurations—with signs and symbols, with patterns, with myth, with language. They deal with perception. Thus the shift from the mechanical to the biological universe will eventually require a new philosophical synthesis. Kant might have called it *Einsicht*, or a *Critique of Pure Perception.*

AFTERWORD:
THE CHALLENGE AHEAD

We cannot yet tell with certainty what the next society and the next economy will look like. We are still in the throes of a transition period. Contrary to what most everybody believes, however, this transition period is remarkably similar to the two transition periods that preceded it during the nineteenth century: the one in the 1830s and 1840s, following the invention of railroads, postal services, telegraph, photography, limited-liability business, and investment banking; and the second one, in the 1870s and 1880s, following the invention of steel making; electric light and electric power; synthetic organic chemicals, sewing machines and washing machines; central heating; the subway; the elevator and with it apartment and office buildings and skyscrapers; the telephone and typewriter and with them the modern office; the business corporation and commercial banking. Both periods were characterized by the paradox of rapidly expanding economy and growing income inequality—the paradox that bedevils us now. And so, while we cannot yet say what the future will be like, we can with very high probability, discern its main and most important features, and some of its main and most important challenges.

The first thing to say may well be that—again contrary to what most everybody believes—it will not be a future of expanding, free markets as we have understood free markets, that is, as markets for the exchange of goods and services. On the contrary, those markets may well shrink, if only because the growth sectors of tomorrow's society are surely going to be two knowledge areas, health care and education, neither of which has ever been, or will ever be, a truly free market. "Free market" tomorrow means flow of information rather than trade. And in that respect, tomorrow will indeed be a worldwide free market. This has serious implications for all institutions, and not only for business. It means, for instance, that every organization everywhere (and not only businesses), will have to be globally competitive.

It also means that the center of gravity, and the center of power, will be the customer. In the last thirty years, the center of power has shifted from the supplier, the manufacturer, to the distributor. In the next thirty years, it will certainly shift to the customer—for the simple reason that the customer now has full access to information worldwide.

We can also anticipate, with very high probability, that the decline in the terms of trade (that is, the purchasing power) of manufacturing, will continue and probably at an accelerated pace. Beginning after World War I, if not in the late nineteenth century, the purchasing power of primary products, especially of farm products in relation to manufactured goods, began to go down sharply. In the twentieth century, it went down at the rate of 1 percent a year compound, which means that by the year 2000 agricultural products bought only one-third of the manufactured goods they had bought in 1900. Beginning in 1960, manufactured goods began to decline in terms of relative purchasing power, that is, in terms of trade, against knowledge goods. Between 1960 and 2000, prices of manufactured goods, adjusted for inflation, fell by almost three-fifths, that is, by 60 percent. At the same time, the prices of the two main knowledge products, health care and education, went up three times as fast as inflation. Manufactured goods by the year 2000 had therefore only about one-fifth of the purchasing power relative to knowledge goods that they had had only forty years earlier.

But the most important certainty is that the next society and economy will have a totally different social complexion. It will be a knowledge society with knowledge workers, the largest single and by far the most expensive part of the workforce. In fact, this has already happened in all developed countries.

And finally, we can say with near-certainty, the challenges we will face in the next economy are management challenges that will have to be tackled by individuals. Governments will be able to help or hinder. But the tasks themselves are not tasks governments can perform. They can only be performed through individual organizations—both business enterprises and nongovernmental nonprofit organizations—and by individuals. Government will not become less pervasive, less powerful, let alone less expensive. It will, however, increasingly depend for its effectiveness on what managers and professionals are doing in and with their own nongovernmental organization, and in and with their own lives.

I hope that *The Essential Drucker* will give the managers, executives, and professionals of tomorrow the understanding of both the society and economy they inherit, and the tools to perform the tasks with which the next society and the next economy will confront them.

—*Peter F. Drucker*
Claremont, California
Spring 2001

INDEX